DESIGNERS' GUIDES TO THE EUROCODES

DESIGNERS' GUIDE TO EUROCODE 2: DESIGN OF CONCRETE STRUCTURES

DESIGNERS' GUIDE TO EN1992-1-1 AND EN1992-1-2 EUROCODE 2: DESIGN OF CONCRETE STRUCTURES DESIGN OF CONCRETE STRUCTURES GENERAL RULES AND RULES FOR BUILDINGS AND STRUCTURAL FIRE DESIGN

Eurocode Designers' Guide Series

Designers' Guide to EN 1990. Eurocode: Basis of Structural Design. H. Gulvanessian, J.-A. Calgaro and M. Holický. 0 7277 3011 8. Published 2002.

Designers' Guide to EN 1994-1-1. Eurocode 4: Design of Composite Steel and Concrete Structures. Part 1.1: General Rules and Rules for Buildings. R. P. Johnson and D. Anderson. 0 7277 3151 3. Published 2004.

Designers' Guide to EN 1997-1. Eurocode 7: Geotechnical Design – General Rules. R. Frank, C. Bauduin, R. Driscoll, M. Kavvadas, N. Krebs Ovesen, T. Orr and B. Schuppener. 0 7277 3154 8. Published 2004.

Designers' Guide to EN 1993-1-1. Eurocode 3: Design of Steel Structures. General Rules and Rules for Buildings. L. Gardner and D. Nethercot. 0 7277 3163 7. Published 2004.

Designers' Guide to EN 1998-1 and EN 1998-5. Eurocode 8: Design Structures for Earthquake Resistance. General Rules, Seismic Actions and Rules for Buildings and Foundations. M. Fardis, E. Carvalho, A. Elnashai, E. Faccioli, P. Pinto and A. Plumier. 0 7277 3348 6. Published 2005.

Designers' Guide to EN 1992-1-1 and EN 1992-1-2. Eurocode 2: Design of Concrete Structures. General Rules and Rules for Buildings and Structural Fire Design. A. W. Beeby and R. S. Narayanan. 0 7277 3105 X. Published 2005.

Designers' Guide to EN 1991-4. Eurocode 1: Actions on Structures. Wind Actions. N. Cook. 0 7277 3152 1. Forthcoming: 2005 (provisional).

Designers' Guide to EN 1996. Eurocode 6: Part 1.1: Design of Masonry Structures. J. Morton. 0 7277 3155 6. Forthcoming: 2005 (provisional).

Designers' Guide to EN 1995-1-1. Eurocode 5: Design of Timber Structures. Common Rules and Rules for Buildings. C. Mettem. 0 7277 3162 9. Forthcoming: 2005 (provisional).

Designers' Guide to EN 1991-1-2, 1992-1-2, 1993-1-2 and EN 1994-1-2. Eurocode 1: Actions on Structures. Eurocode 3: Design of Steel Structures. Eurocode 4: Design of Composite Steel and Concrete Structures. Fire Engineering (Actions on Steel and Composite Structures). Y. Wang, C. Bailey, T. Lennon and D. Moore. 0 7277 3157 2. Forthcoming: 2005 (provisional).

Designers' Guide to EN 1992-2. Eurocode 2: Design of Concrete Structures. Bridges. D. Smith and C. Hendy. 0 7277 3159 9. Forthcoming: 2005 (provisional).

Designers' Guide to EN 1993-2. Eurocode 3: Design of Steel Structures. Bridges. C. Murphy and C. Hendy. 0 7277 3160 2. Forthcoming: 2005 (provisional).

Designers' Guide to EN 1994-2. Eurocode 4: Design of Composite Steel and Concrete Structures. Bridges. R. Johnson and C. Hendy. 0 7277 3161 0. Forthcoming: 2005 (provisional).

Designers' Guide to EN 1991-2, 1991-1-1, 1991-1-3 and 1991-1-5 to 1-7. Eurocode 1: Actions on Structures. Traffic Loads and Other Actions on Bridges. J.-A. Calgaro, M. Tschumi, H. Gulvanessian and N. Shetty. 0 7277 3156 4. Forthcoming: 2005 (provisional).

Designers' Guide to EN 1991-1-1, EN 1991-1-3 and 1991-1-5 to 1-7. Eurocode 1: Actions on Structures. General Rules and Actions on Buildings (not Wind). H. Gulvanessian, J.-A. Calgaro, P. Formichi and G. Harding. 0 7277 3158 0. Forthcoming: 2005 (provisional).

www. eurocodes.co.uk

DESIGNERS' GUIDE TO EUROCODE 2: DESIGN OF CONCRETE STRUCTURES

DESIGNERS' GUIDE TO EN1992-1-1 AND EN1992-1-2 EUROCODE 2: DESIGN OF CONCRETE STRUCTURES DESIGN OF CONCRETE STRUCTURES GENERAL RULES AND RULES FOR BUILDINGS AND STRUCTURAL FIRE DESIGN

A. W. BEEBY and R. S. NARAYANAN

Series editor
H. Gulvanessian

Published by ICE Publishing, 40 Marsh Wall, London E14 9TP
URL: www.icevirtuallibrary.com

Distributors for ICE Publishing Books are
USA: Publishers Storage and Shipping Corp., 46 Development Road, Fitchburg, MA 01420

First published 2005
Reprinted with amendments 2009
Reprinted 2010 and 2011

Eurocodes Expert

Structural Eurocodes offer the opportunity of harmonized design standards for the European construction market and the rest of the world. To achieve this, the construction industry needs to become acquainted with the Eurocodes so that the maximum advantage can be taken of these opportunities

Eurocodes Expert is a new ICE and Thomas Telford initiative set up to assist in creating a greater awareness of the impact and implementation of the Eurocodes within the UK construction industry

Eurocodes Expert provides a range of products and services to aid and support the transition to Eurocodes. For comprehensive and useful information on the adoption of the Eurocodes and their implementation process please visit our website or email eurocodes@thomastelford.com

A catalogue record for this book is available from the British Library

ISBN: 978-0-7277-3105-0

Typeset by Helius, Brighton and Rochester
Printed and bound in Great Britain by CPI Antony Rowe, Chippenham and Eastbourne

Preface

Introduction

EN 1992-1-1 (*General Rules and Rules for Buildings*) and EN 1992-1-2 (*General Rules – Structural Fire Design*) will replace BS 8110 Parts 1 and 2 in the near future. While the broad requirements of EN 1992-1-1 and EN 1992-1-2 are not dissimilar to those in BS 8110, designers will still need to get used to new terminology, set of new documents and the interaction between them.

This guide has been written with the aim of providing practising civil engineers with some insight into the background to EN 1992-1-1 and EN 1992-1-2. The authors have been involved with the evolution of the codes from their ENV (pre-standard) status. The guide starts with a brief outline of the Eurocode system and terminology. The code requirements are illustrated by some local examples. Some design aids are also provided. The guide can be used anywhere in Europe; but it should be noted that the UK values for the Nationally Determined Parameters (set by the UK National Annex at the time of going out to print) have been used in the handbook generally. Some adjustments may be required in this regard when used outside the UK.

All the practical aspects of application of EN 1992-1-1 and EN 1992-1-2 to prestressed concrete design are included in Chapter 11 of this guide. The depth of coverage is limited, but the authors are indebted to Mr Keith Wilson of Faber Maunsell for drafting this chapter.

It is hoped that this guide will facilitate the effective use of Eurocode 2 by designers.

Layout of this guide

All cross-references in this guide to sections, clauses, subclauses, paragraphs, annexes, figures, tables and expressions of EN 1992-1-1 are in *italic type*, which is also used where text from EN 1992-1-1 has been directly reproduced (conversely, quotations from other sources, including other Eurocodes, and cross-references to sections, etc., of this guide, are in roman type). Expressions repeated from EN 1992-1-1 retain their numbering; other expressions have numbers prefixed by D (for Designers' Guide), e.g. equation (D5.1) in Chapter 5. The foregoing also applies to cross-references to EN 1992-1-2, discussed in Chapter 12.

R. S. Narayanan
A. W. Beeby

Contents

CHAPTER 1

Introduction

1.1. Scope

Eurocode 2, *Design of Concrete Structures*, will apply to the design of building and civil engineering structures in plain, reinforced and prestressed concrete. The code has been written in several parts, namely:

- EN 1992-1-1, *General Rules and Rules for Buildings*
- EN 1992-1-2, *General Rules – Structural Fire Design*
- EN 1992-2, *Reinforced and Prestressed Concrete Bridges*
- EN 1992-3, *Liquid and Containment Structures.*

EN 1992-1-1 has been written in such a way that the principles of the code will generally apply to all the parts. The specific rules, which only apply to building structures, are identified as such. Under the CEN (the European standards body) rules, other parts of Eurocode 2 are allowed to identify those clauses in Part 1.1 which do not apply to that part and provide other information that will complement Part 1.1.

This guide is concerned primarily with Part 1.1. Some limited information on Part 1.2 is also provided in Chapter 12.

Part 1.1 covers *in situ* and precast structures using normal-weight or lightweight concrete. It applies to plain, reinforced and prestressed concrete structures. Thus, many of the separate parts of the ENV versions of the code covering the above topics have been brought into one document. Part 1.1 has 12 main chapters and 10 annexes; Part 1-2 has six main chapters and five annexes.

Compliance with the code will satisfy the requirements of the Construction Products Directive in respect of mechanical resistance.

1.2. Layout

The code clauses are set out as Principles and Application Rules. Principles are identified by the letter P following the paragraph number. Application Rules are identified by a number in parentheses.

Principles are general statements and definitions for which there are no alternatives. In addition, they also include some requirements and analytical models for which no alternative is allowed unless specifically stated.

Application Rules are generally accepted methods, which follow the principles and satisfy their requirements. It is permissible to use alternative design rules, provided that it can be demonstrated that they comply with the relevant principles and are at least equivalent with regard to structural safety, serviceability and durability to the rules in the code. This matter should be approached with caution. A narrow interpretation of this requirement will provide no incentive to develop alternative rules. Equivalence could be defined more broadly as meaning that the safety, serviceability and durability that may be expected from using these rules will be sufficient for the purpose. If this is accepted, procedures in the current national codes are, by and large, likely to be acceptable, as the principles are likely to be similar. Clearly, any alternative approach has to be acceptable to regulatory authorities. Provisions of different codes should not be mixed without a thorough appraisal by responsible bodies. It should also be noted that the design cannot be claimed to be wholly in accordance with the Eurocode when an alternative rule is used.

Building regulations will not be harmonized across Europe, and safety in a country remains the prerogative of individual nations. Therefore, in the Eurocode system, some parameters and procedures are left open for national choice. These are referred to as Nationally Determined Parameters (NDPs). These generally relate to safety factors, but not exclusively so. Although at the outset of the conversion of ENVs into ENs, there was a desire by all countries to keep the number of NDPs to a minimum, in practice it has proved difficult to achieve this, and a number of parameters other than safety factors have also become NDPs. The code provides recommended values for all NDPs. Each country is expected to state in their National Annex to the code (which together with the code is likely to form the basis of regulatory control in the country) whether the recommended value is to be changed. Where the UK National Annex alters the recommended value of an NDP, it is identified in this guide.

Chapters are arranged generally by reference to phenomena rather than to the type of element as in UK codes. For example, there are chapters on bending, shear, buckling, etc., but not on beams, slabs or columns. Such a layout is more efficient, as considerable duplication is avoided. It also promotes a better understanding of structural behaviour. However, some exceptions exist such as a chapter on detailing particular member types.

1.3. Related documents

Eurocode 2 refers to a number of CEN standards and codes and some ISO standards where relevant. The following are some of the more important:

- EN 1990, *Basis of Structural Design*
- EN 1991, *Actions on Structures*
- EN 206-1, *Concrete Specification, Performance, Production and Conformity*
- EN 10,080, *Steel Reinforcement of Concrete*
- EN 10,138, *Prestressing Steels*.

Eurocode 2 relies on EN 206-1 for the specification of concrete mixes to ensure durability in various exposure conditions, which are also defined in the same document. In the UK, a complementary standard, BS 8500, has been developed. Part 1 of this British standard is written to assist anyone wishing to specify concrete to BS EN 206-1. Part 2 of this complementary standard contains specification for materials and procedures that are outside of European standardization but within national experience. This part supplements the requirements of BS EN 206-1.

ENV 13670, *Execution of Concrete Structures*, deals with workmanship aspects. It is due to be converted into an EN.

In addition to the above, a number of product standards have been developed for particular products – hollow core units, ribbed elements, etc. While the product standards rely on Eurocode 2 for design matters, some supplementary design rules have been given in some of them.

1.4. Terminology

Generally, the language and terminology employed will be familiar to most engineers. They do not differ from the ENV. EN 1990 defines a number terms, which are applicable to all materials (actions, types of analysis, etc.). In addition, EN 1992-1-1 also defines particular terms specifically applicable to Eurocode 2 A few commonly occurring features are noted below:

- Loads are generally referred to as 'actions' to characterize the generalized format of the code. Actions refer not only to the forces directly applied to the structure but also to imposed deformations, such as temperature effect or settlement. These are referred to as 'indirect actions', and the subscript 'IND' is used to identify this.
- Actions are further subdivided as permanent (G) (dead loads), variable (Q) (live loads or wind loads) and accidental (A). Prestressing (P) is treated as a permanent action in most situations.
- Action effects are the internal forces, stresses, bending moments, shear and deformations caused by actions.
- Characteristic values of any parameter are distinguished by the subscript 'k'. Design values carry the subscript 'd', and take into account partial safety factors.
- The term 'normative' is used for the text of the standards that forms the requirements. The term 'informative' is used only in relation to annexes, which seek to inform rather than require.
- All the formulae and expressions in the code are in terms of the cylinder strength of concrete. It is denoted by f_{ck}. The relationship between the cylinder and cube strengths is set out within the code. The strength class refers to both cylinder and cube strengths, e.g. C25/30 refers to a cylinder strength of 25 N/mm^2 and a corresponding cube strength of 30 N/mm^2. It is not anticipated that the quality control procedures for the production of concrete will switch to cylinder strength where the cube strength is now used (e.g. as in the UK).

CHAPTER 2

Basis of design

2.1. Notation

In this manual symbols have been defined locally where they occur. However, the following is a list of symbols, which occur throughout the document.

$G_{k,j}$	characteristic value of the permanent action j
$G_{k, inf}$	lower characteristic value of a permanent action
$G_{k, sup}$	upper characteristic value of a permanent action
$Q_{k,i}$	characteristic value of the variable action i
A_k	characteristic value of an accidental action
P_k	characteristic value of prestressing force
$\gamma_{G,j}$	partial safety factor for permanent action j for persistent and transient design situations
$\gamma_{GA,j}$	partial safety factor for permanent action j for accidental design situations
$\gamma_{Q,i}$	partial safety factor for variable action i
γ_p	partial safety factor for prestressing force
ψ_0, ψ_1, ψ_2	multipliers for the characteristic values of variable actions to produce combination, frequent and quasi-permanent values, respectively, of variable actions to be used in various verifications
X_K	characteristic value of a material property
γ_M	partial safety factor for the material property, including model uncertainty.

2.2. General

All Eurocodes rely on EN 1990 for the basis of structural design. The provisions of that code are applicable to all materials, and as such only the requirements which are independent of material properties are noted. In the main it provides partial safety factors for actions, including the values that should be used in a load combination.

All Eurocodes are drafted using limit state principles.

A brief résumé of the main requirements of EN 1990 as they affect common designs in concrete is noted below. For a fuller treatment of the subject, reference should be made to *Designers' Guide to EN 1990* in this series of guides to Eurocodes.

2.3. Fundamental requirements

Four basic requirements can be summarized. The structure should be designed and executed in such a way that it will:

(1) during its intended life, with appropriate degrees of reliability and in an economical way

    – sustain all actions and influences that are likely to occur during execution and use
    – remain fit for the intended use
(2) have adequate mechanical resistance, serviceability and durability
(3) in the event of fire, have adequate resistance for the required period of fire exposure
(4) not be damaged by accidents (e.g. explosion, impact and consequences of human error) to an extent disproportionate to the original cause.

According to EN 1990, a design using the partial factors given in its Annex A1 (for actions) and those stated in the material design codes (e.g. Eurocode 2), is likely to lead to a structure with a reliability index β greater than the target value of 3.8 stated in the code for a 50 year reference period (see the note to Table B2 of EN 1990).

2.4. Limit states

Limit states are defined as states beyond which the structure infringes an agreed performance criterion. Two basic groups of limit states to be considered are (1) ultimate limit states and (2) serviceability limit states.

Ultimate limit states are those associated with collapse or failure, and generally govern the strength of the structure or components. They also include loss of equilibrium or stability of the structure as a whole. As the structure will undergo severe deformation prior to reaching collapse conditions (e.g. beams becoming catenaries), for simplicity these states are also regarded as ultimate limit states, although this condition is between serviceability and ultimate limit states; these states are equivalent to collapse, as they will necessitate replacement of the structure or element.

Serviceability limit states generally correspond to conditions of the structure in use. They include deformation, cracking and vibration which:

(1) damage the structure or non-structural elements (finishes, partitions, etc.) or the contents of buildings (such as machinery)
(2) cause discomfort to the occupants of buildings
(3) affect adversely appearance, durability or water and weather tightness.

They will generally govern the stiffness of the structure and the detailing of reinforcement within it.

Figure 2.1 illustrates a typical load–deformation relationship of reinforced-concrete structures and the limit states.

2.5. Actions

2.5.1. Classifications

An action is a direct force (load) applied to a structure or an imposed deformation, such as settlement or temperature effects. The latter is referred to as an indirect action. Accidental actions are caused by unintended events which generally are of short duration and which have a very low probability of occurrence.

The main classification of actions for common design is given in Table 2.1.

2.5.2. Characteristic values of action

Loads vary in time and space. In limit state design, the effects of loads, which are factored suitably, are compared with the resistance of the structure, which is calculated by suitably discounting the material properties. In theory, characteristic values are obtained statistically from existing data. In practice, however, this is very rarely possible, particularly for imposed loads whose nominal values, often specified by the client, are used as characteristic loads. In countries where wind and snow data have been gathered over a period, it will be possible to prescribe a statistically estimated characteristic value.

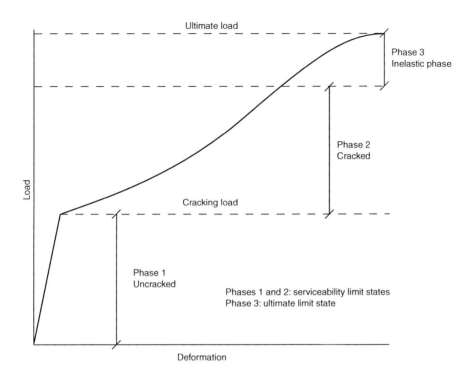

Fig. 2.1. Typical load relationship of reinforced concrete structures and the limit states

Table 2.1. Classification of actions

Permanent action	Variable action	Accidental action
(a) Self-weight of structures, fittings and fixed equipment	(a) Imposed floor loads	(a) Explosions
(b) Prestressing force	(b) Snow loads	(b) Fire
(c) Water and earth loads	(c) Wind loads	(c) Impact from vehicles
(d) Indirect action, e.g. settlement of supports	(d) Indirect action, e.g. temperature effects	

Characteristic values for loads are given EN 1991(*Eurocode 1: Actions on Structures*).

For permanent actions, which vary very little about their mean value (such as weights of materials), the characteristic value corresponds to the mean value. When the variation is likely to be large, e.g. walls or slabs cast against an earth surface with random variations in thickness, or loads imposed by soil fill, upper and lower characteristic values (commonly corresponding to 95th and 5th percentiles) will need to be assessed. These values will apply during the life of the structure. These are denoted by $G_{k, sup}$ and $G_{k, inf}$ respectively.

2.5.3. Design values of actions
The values of actions to be used in design are governed by a number of factors. These include:

(1) The nature of the load. Whether the action is permanent, variable or accidental, as the confidence in the description of each will vary.

(2) The limit state being considered. Clearly, the value of an action governing design must be higher for the ultimate limit state than for serviceability for persistent and transient design situations. Further, under serviceability conditions, loads vary with time, and the design load to be considered could vary substantially. Realistic serviceability loads should be modelled appropriate to the aspect of the behaviour being checked (e.g. deflection, cracking or settlement). For example, creep and settlement are functions of permanent loads only.

(3) The number of variable loads acting simultaneously. Statistically, it is improbable that all loads will act at their full characteristic value at the same time. To allow for this, the characteristic values of actions will need modification.

Consider the case of permanent action (G_k) and one variable action (Q_k) only. For the ultimate limit state the characteristic values should be magnified, and the load may be represented as $\gamma_G G_k + \gamma_Q Q_k$, where the γ factors are the partial safety factors. The values of γ_G and γ_Q will be different, and will be a reflection of the variabilities of the two loads being different. The γ factors account for:

(1) the possibility of unfavourable deviation of the loads from the characteristic values
(2) inaccuracies in the analyses
(3) unforeseen redistribution of stress
(4) variations in the geometry of the structure and its elements, as this affects the determination of the action effects.

Now consider the case of a structure subject to variable actions Q_1 and Q_2 simultaneously. If Q_1 and Q_2 are independent, i.e. the occurrence and magnitude of Q_1 does not depend on the occurrence and magnitude of Q_2 and vice versa, then it would be unrealistic to use $\gamma_{Q,1} Q_{k,1} + \gamma_{Q,2} Q_{k,2}$ as the two loads are unlikely to act at their maximum at the same time. Joint probabilities will need to be considered to ensure that the probability of occurrence of the two loads is the same as that of a single load. It will be more reasonable to consider one load at its maximum in conjunction with a reduced value for the other load. Thus, we have two possibilities:

$$\gamma_{Q,1} Q_{k,1} + \psi_{0,2}(\gamma_{Q,2} Q_{k,2})$$

or

$$\psi_{0,1}(\gamma_{Q,1} Q_{k,1}) + \gamma_{Q,2} Q_{k,2}$$

Multiplication by ψ_0 is said to produce a combination value of the load. It should be noted that the values of γ and ψ_0 vary with each load. EN 1990 provides recommended values, and the UK National Annex will stipulate the values to be used in UK. See also Table 2.3 below. The method of deriving ψ values is outlined in the addenda to ISO 2394:1986. In practice, the designer will not have sufficient information to vary the ψ values in most cases. Table 2.4, below, summarizes the ψ values recommended by the UK National Annex.

The above discussion illustrates the thinking behind the method of combining loads for an ultimate limit state check. Similar logic is applied to the estimation of loads for the different serviceability checks.

Ultimate limit state

(1) **Persistent and transient situations – fundamental combinations**. In the following paragraphs, various generalized combinations of loads are expressed symbolically. It should be noted that the '+' symbol in the expressions does not have the normal mathematical meaning, as the directions of loads could be different. It is best to read it as meaning 'combined with'.

EN 1990 gives three separate sets of load combinations, namely EQU (to check against loss of equilibrium), STR (internal failure of the structure governed by the strength of the construction materials) and GEO (failure of the ground, where the strength of soil provides the significant resistance).

Equilibrium. Equilibrium is verified using the load combination Set A in the code, which is as follows:

$$\gamma_{G,j,sup}G_{k,j,sup} + \gamma_{G,j,inf}G_{k,j,inf} + \gamma_{Q,1}Q_{k,1} + \gamma_{Q,i}\psi_{0,i}Q_{k,i}$$

$\gamma_{G,j,sup}G_{k,j,sup}$ is used when the permanent loads are unfavourable, and $\gamma_{G,j,inf}G_{k,j,inf}$ is used when the permanent actions are favourable. Numerically, $\gamma_{G,j,sup} = 1.1$, $\gamma_{G,j,inf} = 0.9$, and $\gamma_Q = 1.5$ when unfavourable and 0 when favourable.

The above format applies to the verification of the structure as a rigid body (e.g. overturning of retaining walls). A separate verification of the limit state of rupture of structural elements should normally be undertaken using the format given below for strength. In cases where the verification of equilibrium also involves the resistance of the structural member (e.g. overhanging cantilevers), the strength verification given below without the above equilibrium check may be adopted. In such verifications, $\gamma_{G,j,inf} = 1.15$ should be used.

Strength. When a design does not involve geotechnical actions, the strength of elements should be verified using load combination Set B. Two options are given. Either combination (6.10) from EN 1990 or the less favourable of equations (6.10a) and (6.10b) may be used:

$$\gamma_{G,j,sup}G_{k,j,sup} + \gamma_{G,j,inf}G_{k,j,inf} + \gamma_{Q,1}Q_{k,1} + \gamma_{Q,i}\psi_{0,i}Q_{k,i} \tag{6.10}$$

$\gamma_{G,j,sup}G_{k,j,sup}$ is used when the permanent loads are unfavourable, and $\gamma_{G,j,inf}G_{k,j,inf}$ is used when the permanent actions are favourable. Numerically, $\gamma_{G,j,sup} = 1.35$, $\gamma_{G,j,inf} = 1.0$, and $\gamma_Q = 1.5$ when unfavourable and 0 when favourable (EN 1990, UK National Annex).

$$\gamma_{G,j,sup}G_{k,j,sup} + \gamma_{G,j,inf}G_{k,j,inf} + \gamma_{Q,i}\psi_{0,i}Q_{k,i} \tag{6.10a}$$

$$\xi\gamma_{G,j,sup}G_{k,j,sup} + \gamma_{G,j,inf}G_{k,j,inf} + \gamma_{Q,1}Q_{k,1} + \gamma_{Q,i}\psi_{0,i}Q_{k,i} \tag{6.10b}$$

Numerically, $\xi = 0.925$, $\gamma_{G,j,sup} = 1.35$, $\gamma_{G,j,inf} = 1.0$, and $\gamma_Q = 1.5$ when unfavourable and 0 when favourable (EN 1990, UK National Annex).

The above combinations assume that a number of variable actions are present at the same time. $Q_{k,1}$ is the dominant load if it is obvious, otherwise each load is in turn treated as a dominant load and the others as secondary. The dominant load is then combined with the combination value of the secondary loads. Both are multiplied by their respective γ values.

The magnitude of the load resulting from equations (6.10a) and (6.10b) will always be less than that from equation (6.10). The size of the reduction will depend on the ratio $\chi = G_k/(G_k + Q_k)$. Table 2.2 gives the reduction factors for different values of χ.

Now turning to the factors $\gamma_{G,inf}$ and $\gamma_{G,sup}$, it will be noted that the numerical values are different in the verification of equilibrium and that of strength. For instance, in an overhanging cantilever beam, the multiplier for self-weight in the cantilever section will be 1.1 ($\gamma_{G,sup}$) and that in the anchor span will be 0.9 ($\gamma_{G,cnf}$). The possible explanation for $\gamma_{G,sup}$ being 1.1 and not 1.35 as in the strength check is that

(a) the variability in self-weight of the element is unlikely to be large
(b) the factor 1.35 has built into it an allowance for structural performance (which is necessary only for strength checks)
(c) the loading in the cantilever will also generally include variable actions, partial safety factors for which will ensure a reasonable overall safety factor.

When a design involves geotechnical action, a number of approaches are given in EN 1990, and the choice of the method is a Nationally Determined Parameter. In the UK, it will generally be necessary to carry out two separate calculations using the load combinations Set C and Set B and the resistances given in EN 1997. For details, reference should be made to EN 1990 and EN 1997.

(2) **Accidental design situation**. The load combination recommended is

$$G_{k,j,\,\mathrm{sup}} + G_{k,j,\,\mathrm{inf}} + A_\mathrm{d} + \psi_{1,i}Q_{k,1} + \psi_{2,i}Q_{k,i}$$

where A_d is the design value of accidental action, $Q_{k,1}$ is the main variable action accompanying the accidental action and $Q_{k,i}$ are other variable actions.

Accidents are unintended events such as explosions, fire or vehicular impact, which are of short duration and which have a low probability of occurrence. Also, a degree of damage is generally acceptable in the event of an accident. The loading model should attempt to describe the magnitude of other variable loads which are likely to occur in conjunction with the accidental load. Accidents generally occur in structures in use. Therefore, the values of variable actions will be less than those used for the fundamental combination of loads in (1) above. To provide a realistic variable load combining with the accidental load, the variable actions are multiplied by different (and generally lower) ψ factors. Multiplier ψ_1 is applied to the dominant action, and ψ_2 to the others. Where the dominant action is not obvious, each variable action present is in turn treated as dominant. γ_Q for accidental situations is unity.

Multiplication by ψ_1 is said to produce a frequent value of the load, and multiplication by ψ_2 the quasi-permanent value. Numerical values for ψ_1 and ψ_2 are given in EN 1990, and confirmed in the UK National Annex. Table 2.4 below summarizes the values.

Serviceability limit state
(3) **Characteristic combination**.

$$\sum G_{k,j}(+P) + Q_{k,1} + \sum_{i>1} \psi_{0,i}Q_{k,i}$$

This represents a combination of service loads, which can be considered rather infrequent. It might be appropriate for checking states such as micro cracking or possible local non-catastrophic failure of reinforcement leading to large cracks in sections.

Table 2.2. Reduction factors for different values of χ

χ	Load from equation (6.10a) Load from equation (6.10)	Load from equation (6.10b) Load from equation (6.10)
0.0	0.70	**1.00**
0.1	0.73	**0.99**
0.2	0.76	**0.99**
0.3	0.78	**0.98**
0.4	0.81	**0.97**
0.5	0.84	**0.96**
0.6	0.87	**0.96**
0.7	0.90	**0.95**
0.8	**0.93**	0.94
0.9	**0.97**	0.93
1.0	**1.00**	0.93

(1) The reduction factors shown in bold should be used.
(2) The table assumes $\gamma_G = 1.35$, $\gamma_Q = 1.5$, $\psi_0 = 0.7$ and $\xi = 0.925$.

Table 2.3. Partial safety factors for actions in building structures – ultimate limit state (in accordance with the UK National Annex)

Action	Combination	
	Fundamental	Accidental
Permanent actions caused by structural and non-structural components		
Stability check		
Unfavourable	1.10	1.00
Favourable	0.90	1.00
Other checks		
Unfavourable	1.35	1.00
Favourable	1.00	1.00
Variable actions		
Unfavourable	1.50	1.00
Accidental actions	–	1.00

(1) The above apply to persistent and transient design situations.
(2) For accidental design situations $\gamma_{G, A} = 1.0$.
(3) Partial safety factor for the prestressing force (γ_p) is generally 1.0.
(4) For imposed deformation, γ_Q for unfavourable effects is 1.2, when linear methods are used. For non-linear methods, $\gamma_Q = 1.5$.

Table 2.4. ψ values

Variable actions	ψ_0	ψ_1	ψ_2
Imposed loads			
Dwellings	0.7	0.5	0.3
Offices	0.7	0.5	0.3
Shopping and congregation areas	0.7	0.7	0.6
Storage	1.0	0.9	0.8
Parking	0.7	0.7	0.6
Wind loads	0.5	0.2	0.0
Snow loads (for altitudes ≤ 1000 m)	0.7	0.2	0.0

For the purposes of Eurocode 2, the three categories of variable actions in the table should be treated as separate and independent actions.

(4) **Frequent combination.**

$$\sum G_{k,j}(+P) + \psi_{1,1}Q_{k,1} + \sum \psi_{2,i}Q_{k,i} \qquad i > 1$$

This represents a combination that is likely to occur relatively frequently in service conditions, and is used for checking cracking.

(5) **Quasi-permanent combination.**

$$\sum G_{k,j}(+P) + \sum \psi_{2,i}Q_{k,i} \qquad i \geq 1$$

This will provide an estimate of sustained loads on the structure, and will be appropriate for the verification of creep, settlement, etc.

It should be realized that the above combinations describe the magnitude of loads which are likely to be present simultaneously. The actual arrangement of loads in position and direction within the structure to create the most critical effect is a matter of structural analysis (e.g. loading alternate or adjacent spans in continuous beams).

Values of γ and ψ applicable in the UK are given in Tables 2.3 and 2.4.

Examples 2.1–2.4 illustrate the use of the combinations noted above. However, in practice simplified methods given in Section 2.5.4 below are likely to be all that is needed for the majority of structures. Also, in practical examples the dominant loads are likely to be fairly obvious, and therefore the designer will generally not be required to go through all the combinations.

2.5.4. Simplified load combinations
Unlike the ENV version of EN 1992-1-1, EN 1990 does not have simplified load combinations. For normal building structures, expressions (6.10), (6.10a) and (6.10b) for the ultimate limit state may also be represented in a tabular form (Table 2.5).

2.6. Material properties
2.6.1. Characteristic values
A material property is represented by a characteristic value X_k, which in general corresponds to a fractile (commonly 5%) in the statistical distribution of the property, i.e. it is the value below which the chosen percentage of all test results are expected to fall.

Generally, in design, only one (lower) characteristic value will be of interest. However, in some problems such as cracking in concrete, an upper characteristic value may be required, i.e. the value of the property (such as the tensile strength of concrete) above which only a chosen percentage to the values are expected to fall.

2.6.2. Design values
In order to account for the differences between the strength of test specimens of the structural materials and their strength *in situ*, the strength properties will need to be reduced. This is achieved by dividing the characteristic values by partial safety factors for materials (γ_M). Thus the design value $X_d = X_k/\gamma_M$. Uncertainties in the resistance models are also

Table 2.5. Partial safety factors for load combinations in EN 1990 – ultimate limit state

Load combination	Permanent load		Variable load				Prestress
			Imposed				
	Adverse	Beneficial	Adverse	Beneficial	Wind		Prestress
Load combination (6.10)							
Permanent + imposed	1.35	1.00	1.50	0	–		1.00
Permanent + wind	1.35	1.00	–	–	1.50		1.00
Permanent + imposed + wind	1.35	1.00	1.50	0	$1.50 \times 0.5 = 0.75$		1.00
Load combination (6.10a)							
Permanent + imposed	1.35	1.0	$\psi_0^* 1.5$	–	–		1.0
Permanent + wind	1.35	1.0	–	–	$\psi_0 1.5 = 0.9$		1.0
Permanent + imposed + wind	1.35	1.0	$\psi_0^* 1.5$	–	$\psi_0 1.5 = 0.9$		1.0
Load combination (6.10b)							
Permanent + imposed	$\xi 1.35 = 1.25$	$\xi 1.0 = 0.925$	1.5	–	–		1.0
Permanent + wind	$\xi 1.35 = 1.25$	$\xi 1.0 = 0.925$	–	–	1.5		1.0
Permanent + imposed + wind	$\xi 1.35 = 1.25$	$\xi 1.0 = 0.925$	1.5	–	$\psi_0 1.5 = 0.9$		1.0

(1) It is assumed that wind is not the leading action.
(2) ψ_0^* will vary with the use of the building.

Table 2.6. Partial safety factors for material properties[*]

Combination	Concrete, γ_C	Reinforcement and prestressing tendons, γ_S
Fundamental	1.5	1.15
Accidental except fire situations	1.2	1.0
Accidental – fire situations	1.0	1.0

[*] See also UK National Annexes for EN 1992-1-1 and EN 1992-1-2, and the background paper.
(1) These factors apply to the ultimate limit state. For serviceability, $\gamma_M = 1$.
(2) These factors apply if the quality control procedures stipulated in the code are followed. Different values of γ_C may be used if justified by commensurate control procedures.
(3) These factors do not apply to fatigue verification.
(4) $\gamma_S = 1.15$ should be applied to the characteristic strength of 500 MPa.

covered by γ_M. Although not stated in the code, γ_M also accounts for local weaknesses and inaccuracies in the assessment of resistance of the section.

The values of partial safety factors for material properties are shown in Table 2.6.

2.7. Geometric data

The structure is normally described using the nominal values for the geometrical parameters. Variability of these is generally negligible compared with the variability associated with the values of actions and material properties.

In special problems such as buckling and global analyses, geometrical imperfections should be taken into account. The code specifies values for these in the relevant sections. Traditionally, geometrical parameters are modified by factors which are additive.

2.8. Verification

Ultimate limit state
(1) When considering overall stability, it should be verified that the design effects of destabilizing actions are less than the design effects of stabilizing actions.
(2) When considering rupture or excessive deformation of a section, member or connection, it should be verified that the design value of internal force or moment is less than the design value of resistance.
(3) It should be ensured that the structure is not transformed into a mechanism unless actions exceed their design values.

Serviceability limit state
(4) It should be verified that the design effects of actions do not exceed a nominal value or a function of certain design properties of materials; for example, deflection under quasi-permanent loads should be less than span/250, and compression stress under a rare combination of loads should not exceed $0.6f_{ck}$.
(5) In most cases, detailed calculations using various load combinations are unnecessary, as the code stipulates simple compliance rules.

2.9. Durability

As one of the fundamental aims of design is to produce a durable structure, a number of matters will need to be considered early in the design process. These include:

- the use of the structure
- the required performance criteria
- the expected environmental conditions
- the composition, properties and performance of the materials
- the shape of members and the structural detailing
- the quality of workmanship, and level of control
- the particular protective measures
- the maintenance during the intended life.

The environmental conditions should be considered at the design stage to assess their significance in relation to durability and to enable adequate provisions to be made for protection of the materials.

Example 2.1
For the frame shown in Fig. 2.2, identify the various load combinations, to check the overall stability (EQU in EN 1990). Assume office use for this building. (Note that the load combinations for the design of elements could be different.)

(1) Notation:

$G_{k, R}$ characteristic dead load/m (roof)
$G_{k, F}$ characteristic dead load/m (floor)
$Q_{k, r}$ characteristic live load/m (roof)
$Q_{k, F}$ characteristic live load/m (floor)
W_k characteristic wind load/frame at each floor level.

(2) The fundamental load combination to be used is

$$\sum \gamma_{G, j} G_{k, j} + \gamma_{Q, 1} Q_{K, 1} + \sum \gamma_{Q, i} \psi_{0, i} Q_{k, i} \qquad i > 1$$

Although the variability of the permanent actions is unlikely to be significant, when considering stability, a distinction between the favourable and unfavourable effects needs to be made. The values for the different parameters (in accordance with the UK National Annex, Table A1.2(A), Set A) are as follows:

$\gamma_{Ginf} = 0.9$
$\gamma_{G, sup} = 1.1$
$\gamma_Q = 1.5$
ψ_0 (imposed loads – offices) = 0.7
ψ_0 (wind loads) = 0.5

Case 1
Treat the wind load as the dominant load (Fig. 2.3).

Case 2
Treat the imposed load on roof as dominant load (Fig. 2.4).

Case 3
Treat the imposed load on floors as the dominant load (Fig. 2.5).

Note
When the wind loading is reversed, another set of combinations will need to be considered. However, in problems of this type, the designer is likely to arrive at the critical combinations intuitively rather than searching through all the theoretical possibilities.

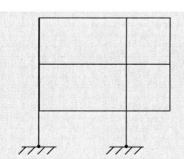

Fig. 2.2. Frame (Example 2.1)

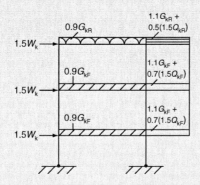

Fig. 2.3. Frame (Example 2.1, case 1)

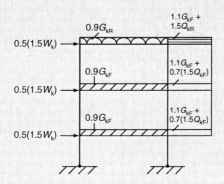

Fig. 2.4. Frame (Example 2.1, case 2)

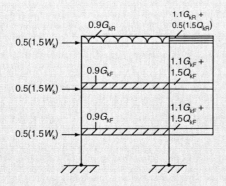

Fig. 2.5. Frame (Example 2.1, case 3)

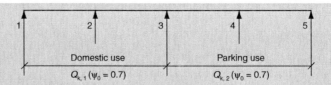

Fig. 2.6. Four-span continuous beam (Example 2.2)

Example 2.2

Identify the various load combinations for the design of a four-span continuous beam for the ultimate limit state (Fig. 2.6). Assume that spans 1–2 and 2–3 are subject to domestic use, and spans 3–4 and 4–5 are subject to parking use.

(1) Notation:

G_k characteristic dead load/m
$Q_{k,1}$ characteristic imposed load/m (domestic use)
$Q_{k,2}$ characteristic imposed load/m (parking use).

(2) The fundamental load combination to be used is

$$\sum \gamma_{G,j} G_{k,j} + \gamma_{Q,1} Q_{k,1} + \sum \gamma_{Q,i} \psi_{0,i} Q_{k,i} \qquad i > 1$$

The same value of self-weight may be applied to all spans, i.e. $1.35G_k$ (EN 1992-1-1, *clause 2.4.3*), as it will produce worse effects than applying $1.00G_k$ throughout.

The load cases to be considered (Fig. 2.7) are:

- alternative spans loaded (EN 1990, UK National Annex, Table A1.2(B), Set B)
- adjacent spans loaded (EN 1990, UK National Annex, Table A1.2(B), note 3).

Example 2.3

For the continuous beam shown in Fig. 2.8, identify the critical load combinations for the design for the ultimate limit state. Assume that the beam is subject to dead and imposed loads and a point load at the end of the cantilever arising from dead loads of the external wall.

(1) Notation:

G_k characteristic dead load/m
Q_k characteristic imposed load/m
P characteristic point load (dead).

(2) Fundamental combinations given in Table A1.2(B) (Set B) of EN 1990 should be used.

Example 2.4

A water tank of depth H m has an operating depth of water h m (Fig. 2.13). Calculate the design lateral loads for the ultimate limit state.

EN 1991-4 (*Actions on Structures, Part 4: Silos and Tanks*) should be used for determining the design loads. The following should be noted:

- The characteristic value of actions corresponds to values that have a probability of 2% that will be exceeded in a reference period of 1 year. In this example, it is assumed that the operational depth h of water has been determined on this basis.

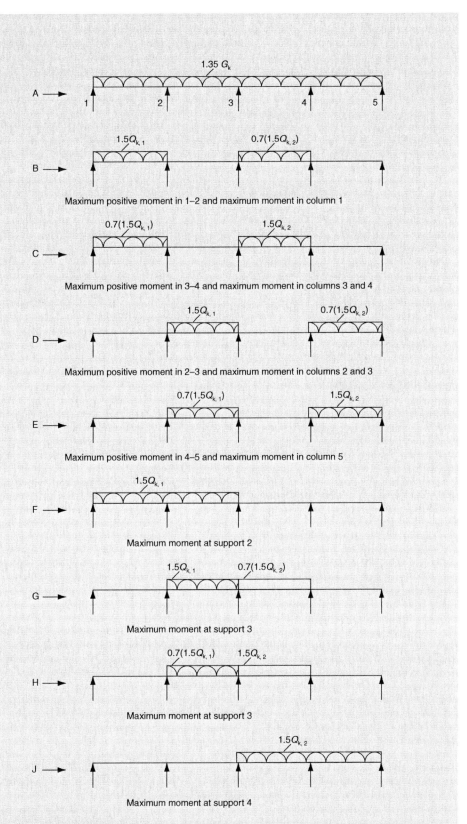

Fig. 2.7. Load cases (Example 2.2)

- This standard also states that the loads on tanks from the stored liquid should be considered when the tank is full. In this example, this condition is treated as an accidental design situation.
- The recommended value of γ_F is 1.2 for the operational condition and 1.0 for accidental situations.

These two cases are shown in Figs 2.14 and 2.15.

Fig. 2.8. Continuous beam (Example 2.3)

Maximum cantilever BM, maximum anchorage of negative steel over 3; also maximum column moment at 3 (see Fig. 2.12)

Fig. 2.9. Continuous beam (Example 2.3, case 1)

Maximum negative moment at column 2

Fig. 2.10. Continuous beam (Example 2.3, case 2)

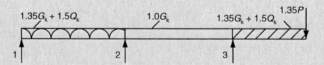

Maximum positive moment 1–2, maximum column moment column 1 and possible maximum moment column 2 (see Fig. 2.12)

Fig. 2.11. Continuous beam (Example 2.3, case 3)

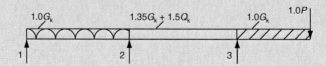

Maximum positive moment 2–3, maximum column moment at 2 (see Fig. 2.11) and maximum column moment at 3 (see Fig. 2.9)

Fig. 2.12. Continuous beam (Example 2.3, case 4)

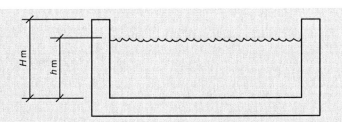

Fig. 2.13. Water tank (Example 2.4)

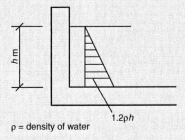

ρ = density of water

Fig. 2.14. Pressure diagram (Example 2.4)

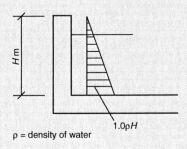

ρ = density of water

Fig. 2.15. Alternative design loading (Example 2.4) – deformation

CHAPTER 3

Analysis

3.1. Introduction

The purpose of analysis is the verification of overall stability and establishment of action effects, i.e. the distribution of internal forces and moments. In turn, this will enable the calculation of stresses, strains, curvature, rotation and displacements. In certain complex structures the type of analysis used (e.g. finite-element analysis) will yield internal stresses and strains and displacements directly.

To carry out the analysis, both the geometry and the behaviour of the structure will need to be idealized.

Commonly, the structure is idealized by considering it as made up of elements depicted in Fig. 3.1.

In terms of the behaviour of the structure, the following methods may be used:

- elastic analysis
- elastic analysis with limited redistribution
- plastic analysis
- non-linear analysis.

The first two models are common for slabs and frames, and plastic analysis is popular in the design of slabs; non-linear analysis is very rarely used in day-to-day design. The above methods, with the exception of plastic analysis, are suitable for both serviceability and ultimate limit states. Plastic methods can be used generally only for ultimate limit state (i.e. strength) design.

In addition to global analysis, local analyses may also be necessary, particularly when the assumption of linear strain distribution does not apply. Examples of this include:

- anchorage zones
- members with significant changes in cross-section, including the vicinity of large holes
- beam–column junctions
- locations adjacent to concentrated loads.

In these cases strut and tie models (a plastic method) are commonly employed to analyse the structures.

3.2. Load cases and combination

In the analysis of the structure, the designer should consider the effects of the realistic combinations of permanent and variable actions. Within each set of combinations (e.g. dead and imposed loads) a number of different arrangements of loads (load cases) throughout the structure (e.g. alternate spans loaded and adjacent spans loaded) will need consideration to identify an envelope of action effects (e.g. bending moment and shear envelopes) to be used in the design of sections.

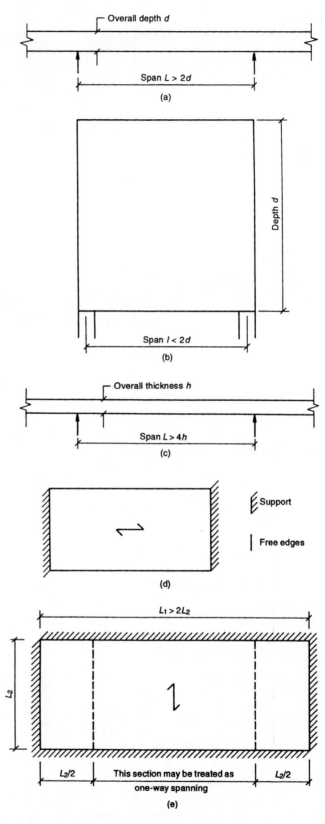

Fig. 3.1. Definition of structural elements for analysis. (a) Beam. (b) Deep beam. (c) Slab. (d, e) One-way spanning slab (subject predominantly to ultimate design load)

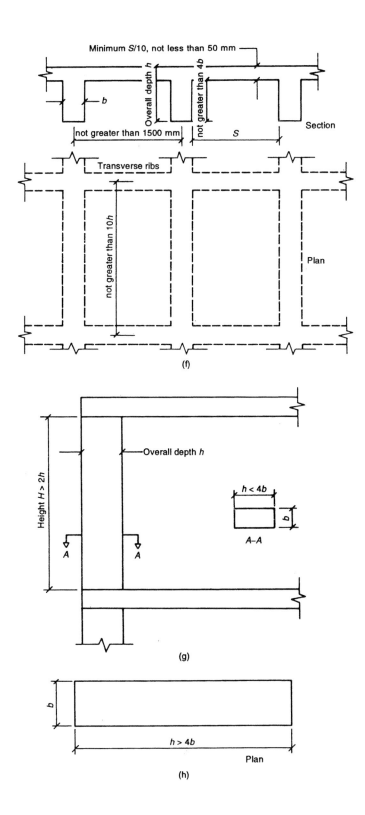

Fig. 3.1. (Contd.) (f) Ribbed and waffle slabs (conditions to be met to allow analysis as solid slabs). (g) Column. (h) Wall

As stated in Chapter 2, EN 1990 provides the magnitude of the design loads to be used when loads are combined. Account is taken of the probability of loads acting together, and values are specifed accordingly.

The EN code for actions (Eurocode 1) specifies the densities of materials (to enable the calculation of permanent actions and surcharges), and values of variable action (such as imposed gravity, wind and snow loads). It also provides information for estimating fire loads in buildings, to enable fire engineering calculations to be carried out.

Although EN 1992-1-1 forms part of a suite of codes including those which specify loads, there is no reason why the Eurocode cannot be used in conjunction with other loading codes. It has been assumed in the Eurocode system that the loads specified in Eurocode 1 are characteristic values with only 5% of values likely to fall above them. Note that for wind loads it is 2%. The definition of the characteristic value will affect the overall reliability.

While the general requirement is that all relevant load cases should be investigated to arrive at the critical conditions for the design of all sections, EN 1992-1-1 permits simplified load arrangements for the design of continuous beams and slabs. The arrangements to be considered are:

Clause 5.1.3

(1) alternate spans loaded with the design variable and permanent loads ($1.35G_k + 1.5Q_k$) and other spans carrying only the design permanent load ($1.35G_k$)
(2) any two adjacent spans carrying the design variable and permanent loads ($1.35G_k + 1.5Q_k$), with all other spans carrying only the design permanent load ($1.35G_k$).

Although not stated, the above arrangements are intended for braced non-sway structures. They may also be used in the case of sway structures, but the following additional load cases involving the total frame will also need to be considered:

(1) all spans loaded with the design permanent loads ($1.35G_k$) and the frame subjected to the design wind load ($1.5W_k$), when W_k is the characteristic wind load
(2) all spans at all floor levels loaded with ($1.35G_k + 1.5Q_k$) and the frame subjected to the design wind load of $1.05W_k$
(3) in sensitive structures (sensitivity to lateral deformation), it may be necessary to consider the effects of wind loading in conjunction with patterned imposed loading through out the frame.

Clause 5.1.3 of EN 1992-1-1 also allows the National Annexes to specify simplification of load arrangements, and the UK National Annex permits the following additional choices.

- For frames:
  - all spans loaded with the maximum design ultimate load ($1.35G_k + 1.5Q_k$)
  - alternate spans loaded with the maximum ultimate load noted above and all other spans loaded with $1.35G_k$.
- For slabs only: a single load case of maximum design load on all spans or panels provided the following conditions are met:
  - in a one-way spanning slab the area of each bay exceeds 30 m^2 (here a bay means a strip across the full width of a structure bounded on the other two sides by lines of supports)
  - the ratio of the characteristic variable load to the characteristic permanent load does not exceed 1.25
  - the characteristic variable load does not exceed 5 kN/m^2.

  The resulting support moments (except those at the supports of cantilevers) should be reduced by 20%, and the span moments adjusted upwards accordingly. No further redistribution should be carried out.

Clause 5.1.1(8)

Clause 5.1.1(8) of EN 1992-1-1 states that in linear elements and slabs subject predominantly to bending, the effects of shear and axial forces on deformation may be neglected, if these are likely to be less than 10%. In practice, the designer need not actually calculate these additional deformations to carry out this check.

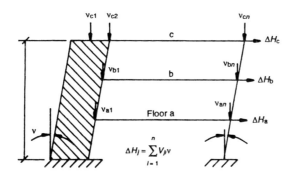

Fig. 3.2. Application of the effective geometrical imperfections: braced structure (number of vertically continuous members = 2)

Deflections are generally of concern only in members with reasonably long spans. In such members, the contribution of shear to the deflections is never significant for members with normal (span/depth) ratios. When the spans are short, EN 1992-1-1 provides alternative design models (e.g. truss or strut and tie) in which deflections are rarely, if ever, a consideration.

The contribution of axial loads to deflections may be neglected if the axial stresses do not exceed $0.08f_{ck}$.

3.3. Imperfections

3.3.1. General

Perfection in buildings exists only in theory; in practice, some degree of imperfection is unavoidable, and designs should recognize this, and ensure that buildings are sufficiently robust to withstand the consequences of such inaccuracies. For example, load-bearing elements may be out of plumb or the dimensional inaccuracies may cause eccentric application of loads. Most codes allow for these by prescribing a notional check for lateral stability. The exact approach adopted to achieve this differs between codes.

EN 1992-1-1 has a number of provisions in this regard, affecting the design of (1) the structure as a whole, (2) some slender elements and (3) elements which transfer forces to bracing members.

3.3.2. Global analysis

For the analysis of the structure as a whole, an arbitrary inclination of the structure $\theta_0 = 1/200$ is prescribed as a basic value. This is then modified for height and for the number of members involved.

The design value will be

$$\theta_i = \theta_0 \alpha_n \alpha_m$$

where

$$\alpha_n = 2/\sqrt{l}$$

where l is the total height of the structure in metres ($0.67 \le \alpha_n \le 1.0$), and

$$\alpha_m = \sqrt{[0.5(1 + 1/m)]}$$

where m is the number of vertically continuous elements in the storeys contributing to the total horizontal force on the floor. This factor recognizes that the degree of imperfection is statistically unlikely to be the same in all the members.

As a result of the inclination, a horizontal component of the vertical loads could be thought of being applied at each floor level, as shown in Figs 3.2 and 3.3. These horizontal

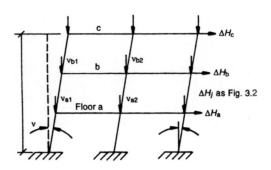

Fig. 3.3. Application of the effective geometrical imperfections: unbraced structure (number of vertically continuous members = 3)

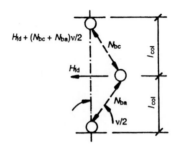

Fig. 3.4. Minimum tie force for perimeter columns

forces should be taken into account in the stability calculation. This is in addition to other design horizontal actions, such as wind.

3.3.3. Design of slender elements
In the design of slender elements, which are prone to fail by buckling (e.g. slender columns), EN 1992-1-1 requires geometrical imperfection to be added to other eccentricities. For example, in the design of the columns, an eccentricity of $\theta_i l_0/2$ is assumed for geometrical imperfection (where l_0 is the effective length of the column).

3.3.4. Members transferring forces to bracing elements
In the design of these elements (such as a floor diagram), a force to account for the possible imperfection should be taken into account in addition to other design actions. This additional force is illustrated in Fig. 3.4. This force need not be taken into account in the design of the bracing element itself.

3.4. Second-order effects
As structures subject to lateral loads deflect, the vertical loads acting on the structure produce additional forces and moments. These are normally referred to as second-order effects. Consider a cantilever column shown in Fig. 3.5. The deflection caused by the horizontal load alone is Δ_1. In this deflected state, the vertical load P will contribute a further bending moment and increase the lateral sway, and the final deflection will be Δ_2. This phenomenon is also commonly referred to as the '$P\Delta$ effect'.

EN 1992-1-1 requires second-order effects to be considered where they may significantly affect the stability of the structure as a whole or the attainment of the ultimate limit state at critical sections.

Clause 5.8.2(6)   In the Application Rules, the code further states that for normal buildings, second-order effects may be neglected if the bending moments caused by them do not increase the first-

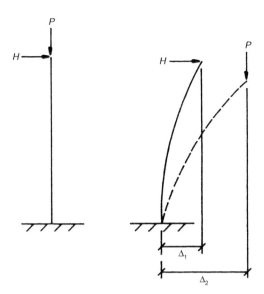

Fig. 3.5. Second-order effect

order bending moments (i.e. bending moments calculated ignoring the effects of displacements) by more than 10%.

Although this would suggest that the designer would first have to check the second-order bending moment before ignoring it, the code provides some simplified criteria, to verify if second-order analysis is required. These tests are summarized in *clause 5.8.3.1* for isolated elements and *clause 5.8.3.3* for structures. These checks essentially ensure that adequate stiffness is provided.

In most practical building structures, second-order effects are unlikely to be significant as serviceability criterion to limit the lateral deflections will ensure that structures are not prone to $P\Delta$ effects.

Clause 5.8.3.1
Clause 5.8.3.3

3.5. Time-dependent effects
The main effects to be considered are creep and shrinkage of concrete and relaxation of prestressing steel. EN 1992-1-1 provides information in *clause 7.2.2*, when such effects need not be considered explicitly.

Clause 7.2.2

3.6. Design by testing
EN 1992-1-1 itself does not provide any useable guidance on this subject. However, there is some guidance in EN 1990. According to EN 1990, testing may be carried out in the following circumstances:

• if adequate calculation models are not available
• if a large number of similar components are to be produced
• to confirm by control checks assumptions made in design.

Annex D of EN 1990 provides further information, particularly regarding the statistics.

Design entirely based on testing is not common in building structures. However, this is an accepted method in some other fields (e.g. pipes). In such work, test programmes should be designed in such a way that an appropriate design strength can be established, which includes proper allowance for the uncertainties covered by partial safety factors in conventional design. It will generally be necessary to establish the influence of material strengths on behaviour and their variability so that a characteristic (and hence design) response can be derived. When testing is carried out on elements which are smaller than the prototype, size effects should be considered in the interpretation of results (e.g. shear strength).

Testing may also be undertaken for other reasons, including:

- appraisal of an existing structure
- to establish data for use in design
- to verify consistency of manufacture or performance of components.

Test methods and procedures will obviously be different in each case. Before planning a test, the precise nature of the information required from the test together with criteria for judging the test should be specified.

3.7. Structural analysis

3.7.1. Elastic analysis with or without redistribution

General

EN 1992-1-1 provides limited guidance on analysis. Elastic analysis remains the most popular method for frame (e.g. moment distribution and slope deflection).

Braced frames may be analysed as a whole frame or may be partitioned into subframes (Fig. 3.6). The subframes may consist of beams at one level with monolithic attachment to the columns. The remote ends may be assumed to be 'fixed' unless a 'pinned' end is more reasonable in particular cases. As a further simplification, beams alone can be considered to be continuous over supports providing no restraint to rotation. Clearly this is more conservative.

In unbraced structures, it is generally necessary to consider the whole structure, particularly when lateral loads are involved. A simplified analysis may be carried out, assuming points of contraflexure at the mid-lengths of beams and columns (Fig. 3.7). However, it should be remembered that this method will be inaccurate if (1) the feet of the column are not fixed and/or (2) the beams and columns are not of similar stiffnesses.

Stiffness parameters

Clause 5.4(2)
Clause 5.4(3)

In the calculation of the stiffness of members it is normally satisfactory to assume a 'mean' value of the modulus of elasticity for concrete and the moment of inertia based on the uncracked gross cross-section of the member. However, when computing the effects of deformation, shrinkage and settlement reduced stiffness corresponding to cracked cross section should be used.

Effective spans

Calculations are performed using effective spans, which are defined below. The principle is to identify approximately the location of the line of reaction of the support.

Clause 5.3.2.2
$$l_{eff} = l_n + a_1 + a_2$$

where l_{eff} is the effective span, l_n is the distance between the faces of supports, and a_1 and a_2 are the distances from the faces of the supports to the line of the effective reaction at the two ends of the member.

Typical conditions are considered below.

- Case 1: intermediate support over which member is continuous (Fig. 3.8).
- Case 2: monolithic end support (Fig. 3.9). a_j is the lesser of half the width of support or half the overall depth of number.
- Case 3: discontinuous end support (Fig. 3.10). a_j is the lesser of half the width of support or third of the overall depth of member.

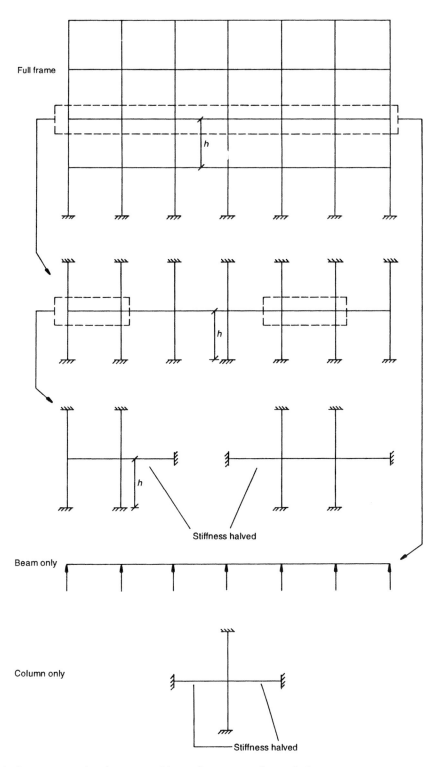

Fig. 3.6. Partitioning of multi-storeyed braced structures for analysis

- Case 4: discontinuous end support on bearings (Fig. 3.11). a_j is the distance from the face of the support to the centre line of the bearing.
- Case 5: isolated cantilever. $a_j = 0$, i.e. the effective span is the length of the cantilever from the face of the support.

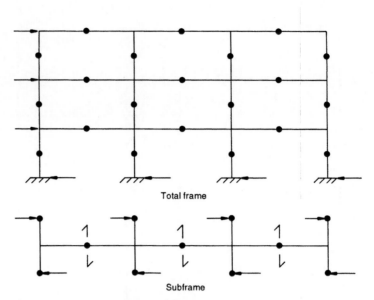

Fig. 3.7. Simplified model for the analysis of unbraced structures

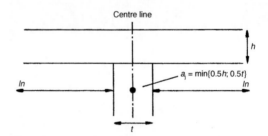

Fig. 3.8. Intermediate support over which member is continuous

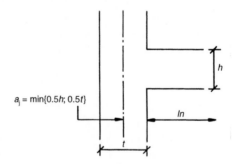

Fig. 3.9. Monolithic end support

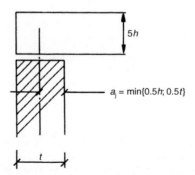

Fig. 3.10. Discontinuous end support

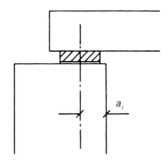

Fig. 3.11. Discontinuous end support on bearings

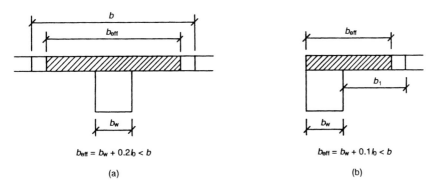

Fig. 3.12. Effective flange width: (a) T beam; (b) L beam. b is the beam spacing and b_1 is half the clear spacing between beams

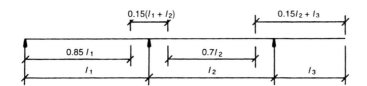

Fig. 3.13. Definition of l_0 for the calculation of the effective flange width

Effective width of flanges
As a result of 'shear lag', the stress in the parts of a wide flange distant from the web would be much less than that at the flange–web junction. The calculation of the variation is a complex mathematical problem. Therefore, codes of practice allow approximations by which an 'effective' width can calculated. A uniform distribution of stress is assumed over the effective width.

For building structures the effective widths shown in Fig. 3.12 may be used.

l_0 in Fig. 3.13 is the distance between points of contraflexure.

Note that:

Clause 5.3.2.1

(1) The length of the cantilever should be less than half the adjacent span.
(2) The ratio of adjacent spans (other than the cantilever) should be between 0.67 and 1.5.

In analysis, EN 1992-1-1 permits the use of a constant flange width throughout the span. This should take the value applicable to span sections. This applies only to elastic design with or without redistribution and not when more rigorous non-linear methods are adopted.

Redistribution of moments
The moment–curvature response of a true elasto-plastic material will be typically as shown at Fig. 3.14. The long plateau after M_p is reached implies a large rotation capacity. Consider a

continuous beam made of such a material and loaded as shown in Fig. 3.15(a). When a bending moment in a critical section (usually at a support) reaches M_p, a plastic hinge is said to be formed. The structure will be able to withstand further increases in loading until sufficient plastic hinges form to turn the structure into a mechanism.

The bending moments at 'failure' will be as shown in Fig. 3.15(c). Had the beam been elastic, then the bending moment will be as shown in Fig. 3.15(b). Comparing these two figures, it can be seen that the elastic bending moment at the support has been reduced from $(3/16)P_uL$ to $(1/6)P_uL$, or redistributed by about 11%. It should also be noted that the span moment has increased by about 7%.

The process illustrated is plastic analysis. Clearly, to exploit plasticity fully, the material must possess adequate ductility (rotation capacity). Concrete has only limited capacity in this regard. The moment redistribution procedure is an allowance for the plastic behaviour without carrying out plastic hinge analysis. Indirectly, it also ensures that the yield of sections under service loads and large uncontrolled deflections are avoided.

EN 1992-1-1 allows moment redistribution up to 30% in braced structures. No redistribution is permitted in sway frames or in situations where rotation capacity cannot be defined with confidence.

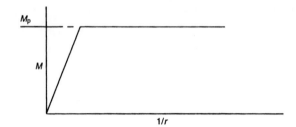

Fig. 3.14. Moment–curvature: idealization

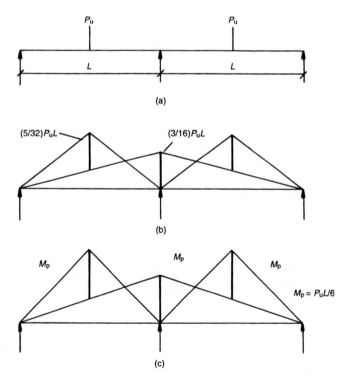

Fig. 3.15. (a, b) Bending moment of an elastic beam at failure. (c) Bending moment of an elasto-plastic beam at failure

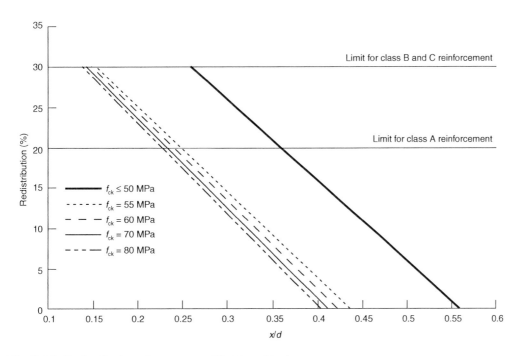

Fig. 3.16. Redistribution of moment and limiting x/d values

When redistribution is carried out, it is essential to maintain equilibrium between the applied loads and the resulting distribution of bending moments. Thus, where support moments are reduced, the moments in the adjacent spans will need to be increased to maintain equilibrium for that particular arrangement of loads.

In EN 1992-1-1 the limit to the amount of redistribution is related to the ductility characteristics of the reinforcement used:

- class B or class C reinforcement – 30%
- class A reinforcement – 20%.

In the case of continuous beams or slabs, subject predominantly to flexure and in which the ratio of the lengths of the adjacent spans lie in the range 0.5 to 2.0, the code permits redistribution to be carried out without the rotation being explicitly checked provided the following expressions are satisfied (see the UK National Annex).

- For $f_{ck} \leq 50$ MPa:

  $\delta \geq 0.4 + [0.6 + (0.0014/\varepsilon_{cu2})]x/d$

- For $f_{ck} > 50$ MPa

  $\delta \geq 0.4 + [0.6 + (0.0014/\varepsilon_{cu2})]x/d$

where δ is the ratio of the redistributed moment to the moment before redistribution, x is the depth of the neutral axis at the ultimate limit state after redistribution, and d is the effective depth. FN values of ε_{cu2} refer to Table 4.1.

These requirements are presented graphically in Fig. 3.16. Table 3.1 gives the x/d values for different values of δ.

It should be noted that the above limitations try to ensure sufficiently ductile behaviour. Higher-strength concrete tends to be brittle, and hence the more onerous limits on x/d. x/d also increases with the steel percentage, and therefore indirectly affects the ductility of the section.

3.7.2. Plastic analysis

Clause 5.6.2(2)

Apart from moment redistribution, EN 1992-1-1 allows the use of plastic analysis without any direct check on rotation capacity, provided certain conditions are satisfied:

(1) The area of reinforcement at any location in any direction should not exceed a value corresponding to $x_u/d = 0.25$ for concrete strength $f_{ck} \leq 50$ MPa and $x_u/d = 0.15$ for concrete strength $f_{ck} \geq 55$ MPa.
(2) Reinforcing steel should be class B or C.
(3) The ratio of the moments over continuous edges to the moments in the span should be between 0.5 and 2.0.

The first of the conditions can be expressed as a reinforcement percentage for a balanced section, i.e. when both concrete and reinforcement have reached their limiting strains (Fig. 3.17):

$$F_c = F_s$$

$$F_c = 0.8x(0.85f_{ck}/1.5)b$$

Substituting $x = 0.25d$:

$$F_c = 0.113bdf_{ck}$$

$$F_s = A_s(0.87f_y)$$

Substituting $A_s = (\rho/100)bd$ and equating F_c and F_s, we obtain

$$\rho = 12.99f_{ck}/f_y$$

Table 3.2 gives the values of ρ for different combinations of f_{ck} and $f_y = 500$ MPa. Note that the stress block is a function of f_{ck} for concrete strength higher than 50 MPa.

Table 3.1. Limiting neutral axis depths and redistribution of moment

	x/d				
δ	$f_{ck} \leq 50$ MPa	$f_{ck} = 55$ MPa	$f_{ck} = 60$ MPa	$f_{ck} = 70$ MPa	$f_{ck} = 80, 90$ MPa
0.70	0.260	0.152	0.148	0.143	0.140
0.80	0.360	0.247	0.240	0.232	0.228
0.85	0.410	0.295	0.286	0.277	0.272
0.90	0.460	0.342	0.332	0.322	0.316

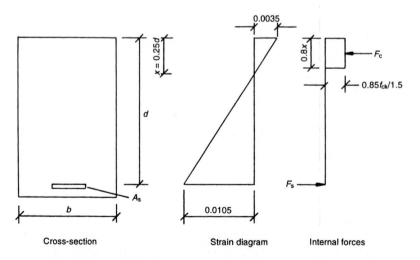

Fig. 3.17. Strain and stress conditions in a balanced section

Table 3.2. Maximum values of $\rho = 100A_s/bd$ for plastic analysis using reinforcement with $f_{yk} = 500$ MPa

f_{ck}	ρ
20	0.52
25	0.65
30	0.78
35	0.91
40	1.04
45	1.17
50	1.30
60	1.00
70	1.09
80	1.13
90	1.16

The most common plastic method used in practice is the yield line analysis for slabs. For details of the method, standard textbooks should be consulted. In conjunction with plastic analysis, an appraisal of the behaviour of the structure at serviceability should be undertaken.

3.7.3. Non-linear analysis

This takes into account the non-linear deformation properties of reinforced-concrete sections. This method is seldom used in practical design as it is complex, requiring a computer and prior knowledge of reinforcement details throughout the structures. It could be useful in appraising the capacity of existing structures or when a large repetition of a particular structure is considered (e.g. precast frames).

The method involves plastic hinge analysis (see redistribution of moments, p. 31), which includes explicit calculation of rotations at hinges. In this method, 'failure' is reached (1) where limiting rotation occurs at a plastic hinge or (2) when sufficient hinges form to render the structure into a mechanism, whichever occurs first. A hinge is said to have formed when the steel starts to yield.

Consider an encastre beam, as shown in Fig. 3.18. The procedure to predict the ultimate load capacity may be as follows:

(1) Knowing the reinforcement, section properties and concrete grade, calculate the M_{yk} at A and C. M_{yk} is the bending moment which produces a stress f_{yk} in the tension reinforcement. Also, calculate M_{yd} by applying γ_m.
(2) Calculate the loading which will produce an elastic bending moment at A and B equal to M_{yk}.
(3) Increase the loading in stages. At each stage calculate (a) the rotation of the hinges at A and B by integrating the curvatures of the beam between hinges (this will require the beam to be divided into a number of sections. At each section the curvature should be calculated using *equation (7.18)* of EN 1992-1-1); (b) the bending moment at C; and (c) the strains in steel and concrete at A, B and C.
(4) Compare the rotation at A and B with the limiting rotation given in *Fig. 5.6N* in EN 1992-1-1.
(5) Failure is reached when either the limiting rotation is reached at A and B or M_{yk} is reached at C.
(6) Calculate the load corresponding to 'failure' as noted above.

Note that the rotations are calculated using mean values of mechanical properties whereas the strengths are calculated using design properties, i.e. γ_m is used.

The above procedure is set out to clarify the steps involved in a relatively simple example. It must be clear that the analysis of a continuous beam will be fairly complex even with a computer. Also, a number of boundary conditions will need to be imposed in the analysis. In practice, therefore, this procedure is rarely used, and elastic analysis with moment redistribution is preferred.

3.7.4. Strut-and-tie models

Strut-and-tie models utilize the lower-bound theorem of plasticity, which can be summarized as follows: for a structure under a given system of external loads, if a stress distribution throughout the structure can be found such that (1) all conditions of equilibrium are satisfied and (2) the yield condition is not violated anywhere, then the structure is safe under the given system of external loads. This approach particularly simplifies the analysis of the parts of the structure where linear distribution of strain is not valid. Typical areas of application are noted in Section 3.1 above.

Typical models are shown in Fig. 3.19. As can be seen, the structure is thought of as comprising notional concrete struts and reinforcement ties. Occasionally, concrete ties may also be considered (e.g. slabs without stirrups; anchorages without transverse reinforcement). While there might appear to be infinite freedom to choose the orientation of struts and ties, this is not so. Concrete has only a limited plastic deformation capacity; therefore, the model has to be chosen with care to ensure that the deformation capacity is not exceeded at any point before the assumed state of stress is reached in the structure. Generally, it will be safe to model the struts and ties in such a way that they closely follow the 'stress paths' indicated by the theory of elasticity. The angles between the struts and ties should generally be greater than 45° in order to avoid incompatibility problems. Where several models are possible, the 'correct' model is the one with the least number of internal members and least deformation. In this context, the deformation of the struts may be neglected, and the model optimized by minimizing the expression

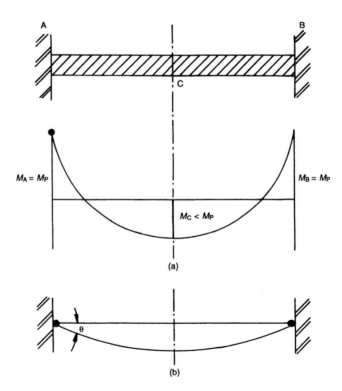

Fig. 3.18. Plastic analysis of an encastre beam: (a) elastic moments when the first hinge(s) form at A and B; (b) rotation θ at the first hinge when the load is increased

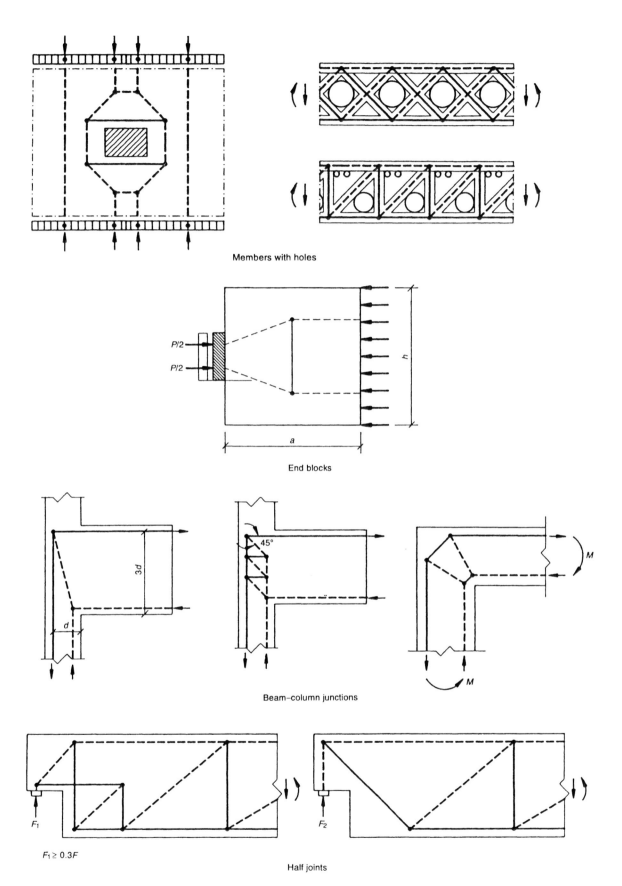

Fig. 3.19. Typical strut-and-tie models

$$\sum F_i l_i \varepsilon_{mi}$$

where F_i is the force in tie, l_i is the length of tie and ε_{mi} is the mean strain in tie i.

Having idealized the structure as struts and ties, it is then a simple matter to arrive at the forces in them based on equilibrium with external loads. Stresses in the struts and ties should be verified including those at nodes, where a number of members meet.

Limiting stresses are as follows:

(1) in ties, f_{yd}
(2) in struts with no transverse tension, $\sigma_{Rd, max} = f_{cd}$
(3) in struts with transverse tension, $\sigma_{Rd, max} = 0.6\nu'f_{cd}$, where $\nu' = 1 - f_{ck}/250$
(4) in nodes where no ties are anchored, $\sigma_{Rd, max} = 1.0\nu'f_{cd}$

Clause 6.5.2
Clause 6.5.4

(5) in compression–tension nodes, where ties are anchored in one direction only, $\sigma_{Rd, max} = 0.85\nu'f_{cd}$
(6) in compression–tension nodes, where ties are anchored in more than one direction, $\sigma_{Rd, max} = 0.75\nu'f_{cd}$
(7) where a load distributed uniformly over an area A_{c0} is dispersed to a larger area A_{c1} (which is concentric to A_{c0}), the applied load F_{Rdu} should be limited as $F_{Rdu} = A_{c0}f_{cd}(A_{c1}/A_{c0})^{0.5}$ but limited to $3.0A_{c0}f_{cd}$.

In general, if the above limits are observed, the 'node' regions are unlikely to be critical. Schlaich and Schafer[1] recommend a procedure for checking the node regions.

Bearing in mind that strut and tie models fall under plastic analysis, it is interesting to note that the code does not impose any conditions similar to those noted in Section 3.7.2. This is an inconsistency. It is therefore important to follow the theory of elasticity fairly closely in choosing the model as already discussed above.

3.8. Design aids and simplifications

3.8.1. One-way spanning slabs and continuous beams

The bending moment coefficients shown in Fig. 3.20 may be used for three or more equal spans subject to equal uniformly distributed loads. This assumes supports which do not offer rotational restraint. The coefficients for locations 3–6 may be used (without sacrificing too much economy) even when continuity with internal columns is to be taken into account. However, continuity with external columns has a significant effect on the bending moments at locations 1 and 2. The following modifications may be made, depending on the ratio of the total column wall stiffness to the slab stiffness.

It should be noted that the values of bending moments in Fig. 3.20 for variable loads are the maximum values obtained by a linear analysis considering the cases of loading alternate and adjacent spans. It will not, therefore, be possible to redistribute the bending moments, as the loading case applicable will vary with each span.

A further approximation may be made if: (1) the characteristic variable load does not exceed the characteristic permanent load; (2) loads are predominantly uniformly distributed over three or more spans; and (3) variations in span lengths do not exceed 15% of the longest span. Under these conditions the bending moments and shear forces may be obtained using Table 3.3.

Table 3.3 assumes that the ratio of the stiffness of the end columns to that of the slab (or beam) does not exceed 1.0. No redistribution of moments is permitted to the bending moments obtained using the above table.

3.8.2. Two-way spanning slabs

General

For rectangular slabs with standard edge conditions and subject to uniformly distributed loads, normally the bending moments are obtained using tabulated coefficients. Such coefficients are provided later in this section.

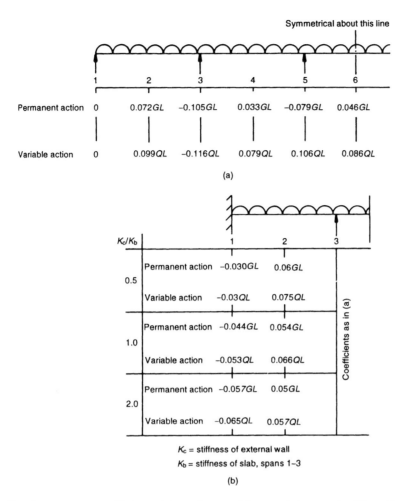

Fig. 3.20. Bending moment coefficients: (a) continuous slab on point supports; (b) continuous slabs with continuity at external columns

Table 3.3. Ultimate bending moment and shear forces in one-way spanning slabs (continuous beams and flat slabs)

	End support and end span				At first interior support	Middle of interior spans	Interior supports
	Simple		Monolithic				
	At outer support	Near middle of end span	At outer support	Near middle of end span			
Moment	0	0.086*Fl*	−0.040*Fl*	0.075*Fl*	−0.086*Fl*	−0.063*Fl*	−0.063*Fl*
Shear	0.40*F*	−	0.46*F*	−	0.6*F*	−	0.50*F*

F is the total design ultimate load ($1.35G_k + 1.50Q_k$); *l* is the effective span.

For slabs with irregular plan shapes and slabs subject to a combination of point loads and distributed loads, Johansen's yield line analysis and the Hillerborg strip method provide powerful methods for strength calculations (e.g. see Johansen,[2] *CEB Bulletin 35*[3] and Wood and Armer[4]).

Simply supported slabs

Where (1) corners of slabs are free to lift and (2) no provision is made to resist forces at the corners, the maximum moments per unit width are given by the following expressions:

$$M_{sx} = \text{bending moment in strips with span } l_x$$
$$= \alpha_{sx} q l_x^2$$

$$M_{sy} = \text{bending moment in strips with span } l_y$$
$$= \alpha_{sy} q l_x^2$$

where l_x is the shorter span of the panel, l_y is the longer span of the panel and q is the design ultimate load per unit area. Values of α_{sx} and α_{sy} are given in Table 3.4 for different ratios of l_y and l_x, where l_y is the longer span.

Rectangular panels with restrained edges

Where corners are (1) prevented from lifting and (2) reinforced to resist torsion, the maximum bending moments per unit width are given by the following expressions:

$$M_{sx} = \beta_{sx} q l_x^2$$
$$M_{sy} = \beta_{sy} q l_x^2$$

where M_{sx} is the maximum design moment either over supports or at midspan on strips with span l_x, M_{sy} is the maximum design moment either over supports or at midspan on strips with span l_y, q is the design ultimate load per unit area, l_x is the shorter span and l_y is the longer span.

For different ratios of the longer to shorter spans (l_y/l_x), values of β_{sx} and β_{sy} are tabulated in Table 3.5 for different edge conditions. Conditions for the use of the tabular value are:

(1) The table is based on the yield line (i.e. plastic analysis). Therefore, conditions (1)–(3) in Section 3.7.2 should be observed.
(2) The permanent and variable loads in adjacent panels should be approximately the same as on the panel being designed.
(3) The span of the adjacent panels in the direction perpendicular to the line of the common support should be approximately the same as the span of the panel being designed.
(4) Corners at the junction of simply supported edges should be reinforced over the zone shown in Fig. 3.21.
(5) The panel should observe the detailing rules in Chapter 10 of this guide.

In the corner area shown:

(1) provide top and bottom reinforcement
(2) in each layer provide bars parallel to the slab edges

Table 3.4. Bending moment coefficients for slabs spanning in two directions at right angles, simply supported on four sides

l_y/l_x	α_{sx}	α_{sy}
1.0	0.062	0.062
1.1	0.074	0.061
1.2	0.084	0.059
1.3	0.093	0.055
1.4	0.099	0.051
1.5	0.104	0.046
1.75	0.113	0.037
2.0	0.118	0.029

Table 3.5. Bending moment coefficients for rectangular panels supported on four sides with provision for torsion at corners

Type of panel and moments considered	Short span coefficients β_{sx}								Long span coefficients β_{sy}, for all values of l_y/l_x
l_y/l_x:	1.0	1.1	1.2	1.3	1.4	1.5	1.75	2.0	
Interior panels									
Negative moment at continuous edge	0.031	0.037	0.042	0.046	0.050	0.053	0.059	0.063	0.032
Positive moment at mid-span	0.024	0.028	0.032	0.035	0.037	0.040	0.044	0.048	0.024
One short edge discontinuous									
Negative moment at continuous edge	0.039	0.044	0.048	0.052	0.055	.0.58	0.063	0.067	0.037
Positive moment at mid-span	0.029	0.033	0.036	0.039	0.041	0.043	0.047	0.050	0.028
One long edge discontinuous									
Negative moment at continuous edge	0.039	0.049	0.056	0.062	0.068	0.073	0.082	0.089	0.037
Positive moment at mid-span	0.030	0.036	0.042	0.047	0.051	0.055	0.062	0.067	0.028
Two adjacent edges discontinuous									
Negative moment at continuous edge	0.047	0.056	0.063	0.069	0.074	0.078	0.087	0.093	0.045
Positive moment at mid-span	0.036	0.042	0.047	0.051	0.055	0.059	0.065	0.070	0.034
Two short edges discontinuous									
Negative moment at continuous edge	0.046	0.050	0.054	0.057	0.060	0.062	0.067	0.070	–
Positive moment at mid-span	0.034	0.038	0.040	0.043	0.045	0.047	0.050	0.053	0.034
Two long edges discontinuous									
Negative moment at continuous edge	–	–	–	–	–	–	–	–	0.045
Positive moment at mid-span	0.034	0.046	0.056	0.065	0.072	0.078	0.091	0.100	0.034
Three edges discontinuous (one long edge continuous)									
Negative moment at continuous edge	0.057	0.065	0.071	0.076	0.081	0.084	0.092	0.098	–
Positive moment at mid-span	0.043	0.048	0.053	0.057	0.060	0.063	0.069	0.074	0.044
Three edges discontinuous (one short edge continuous)									
Negative moment at continuous edge	–	–	–	–	–	–	–	–	0.058
Positive moment at mid-span	0.042	0.054	0.063	0.071	0.078	0.084	0.096	0.105	0.044
Four edges discontinuous									
Positive moment at mid-span	0.055	0.065	0.074	0.081	0.087	0.092	0.103	0.111	0.056

(3) in each of the four layers the area of reinforcement should be equal to 75% of the reinforcement required for the maximum span moment

(4) the area of reinforcement in (3) above may be halved if one edge of the slab in the corner is continuous.

Unequal edge conditions in adjacent panels

M_{-1} and M_{-2} are the support moments for panels 1 and 2, respectively, and M_{+1} and M_{+2} are the span moments for panels 1 and 2, respectively.

In some cases, the bending moments at a common support, obtained by considering the two adjacent panels in isolation, may differ significantly (say by 10%), because of the differing edge condition at the far supports or differing span lengths or loading.

Consider panels 1 and 2 in Fig. 3.22. As the support on grid A for panel 1 is discontinuous and support on grid C for panel 2 is continuous, the moments for panels 1 and 2 for the support on grid B could be significantly different. In these circumstances, the

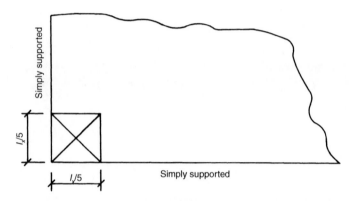

Fig. 3.21. Corner reinforcement: two-way spanning slabs

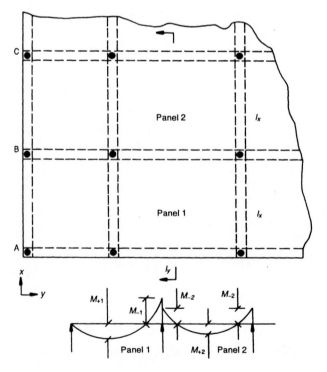

Fig. 3.22. Two-way spanning slabs: unequal edge condition in adjacent panels

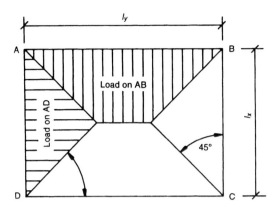

Fig. 3.23. Loads on beams supporting two-way spanning slabs

slab may be reinforced throughout for the worst case span and support moments. However, this might be uneconomic in some cases. In such cases, the following distribution procedure may be used:

(1) Obtain the support moments for panels 1 and 2 from Table 3.6. Treating M_{-1} and M_{-2} as fixed end moments, the moments may be distributed in proportion to the stiffnesses of span l_x in panels 1 and 2. Thus, a revised bending moment M'_{-B} may be obtained for support over B.
(2) The span moments in panels 1 and 2 should be recalculated as follows:

$$M'_{+1} = (M_{-1} + M_{+1}) - M^1_{-B}$$

$$M'_{+2} = (M_{-2} + M_{-2} + M_{+2}) - M'_{-B} - M_{-2}$$

(Note that this assumes that the final moment over C is M_{-2}.)
(3) For curtailment of reinforcement, the point of contraflexure may be obtained by assuming a parabolic distribution of moments in each panel.

Loads on supporting beams
Loads on supporting beams may be obtained using either Fig. 3.23 or Table 3.6.
    Note that:

(1) the reactions shown apply when all edges are continuous (or discontinuous)
(2) when one edge is discontinuous, the reactions on all continuous edges should be increased by 10% and the reaction on the discontinuous edge may be reduced by 20%
(3) when adjacent edges are discontinuous, the reactions should be adjusted for elastic shear, considering each span separately.

3.8.3. Flat slabs
Flat slab structures are defined as slabs (solid or coffered) supported on point supports. Unlike two-way spanning slabs on line supports, flat slabs can fail by yield lines in either of the two orthogonal directions (Fig. 3.24). For this reason, flat slabs must be capable of carrying the total load on the panel in **each** direction.

Methods of analyses
Many recognized methods are available. These include:

(1) the equivalent frame method
(2) simplified coefficients
(3) yield line analysis
(4) grillage analysis.

In this guide, only the first two of these methods will be considered. For other approaches specialist literature should be consulted (e.g. *CIRIA Report 110*[5]).

Equivalent frame method
Division into frames
The structure may be divided in two orthogonal directions into frames consisting of columns and strips of slabs acting as 'beams'. The width of the slab to be used for assessing the stiffness depends on the aspect ratio of the panels and whether the loading is vertical or horizontal.

Vertical loading. When the aspect ratio is less than 2 the width may be taken as the distance between the centre lines of the adjacent panels. For aspect ratios greater than 2, the width may be taken as the distance between the centre lines of the adjacent panels when considering bending in the direction of longer length spans of the panel and twice this value for bending in the perpendicular direction. (See Fig. 3.25.)

The width of beams for frame analysis is as follows:

Table 3.6. Shear force coefficients for uniformly loaded rectangular panels supported on four sides with provision for torsion at corners

Type of panel and location	β_{vx} for values of γ_y/γ_x								β_{vy}
	1.0	1.1	1.2	1.3	1.4	1.5	1.75	2.0	
Four edges continuous									
Continuous edge	0.33	0.36	0.39	0.41	0.43	0.45	0.48	0.50	0.33
One short edge discontinuous									
Continuous edge	0.36	0.39	0.42	0.44	0.45	0.47	0.50	0.52	0.36
Discontinuous edge	–	–	–	–	–	–	–	–	0.24
One long edge discontinuous									
Continuous edge	0.36	0.40	0.44	0.47	0.49	0.51	0.55	0.59	0.36
Discontinuous edge	0.24	0.27	0.29	0.31	0.32	0.34	0.36	0.38	–
Two adjacent edges discontinuous									
Continuous edge	0.40	0.44	0.47	0.50	0.52	0.54	0.57	0.60	0.40
Discontinuous edge	0.26	0.29	0.31	0.33	0.34	0.35	0.38	0.40	0.26
Two short edges discontinuous									
Continuous edge	0.26	0.30	0.33	0.36	0.38	0.40	0.44	0.47	–
Discontinuous edge	–	–	–	–	–	–	–	–	0.40
Two long edges discontinuous									
Continuous edge	–	–	–	–	–	–	–	–	0.40
Discontinuous edge	0.26	0.30	0.33	0.36	0.38	0.40	0.44	0.47	–
Three edges discontinuous (one long edge continuous)									
Continuous edge	0.45	0.48	0.51	0.53	0.55	0.57	0.60	0.63	–
Discontinuous edge	0.30	0.32	0.34	0.35	0.36	0.37	0.39	0.41	0.29
Three edges discontinuous (one short edge continuous)									
Continuous edge	–	–	–	–	–	–	–	–	0.45
Discontinuous edge	0.29	0.33	0.36	0.38	0.40	0.42	0.45	0.48	0.30
Four edges discontinuous									
Discontinuous edge	0.33	0.36	0.39	0.41	0.43	0.45	0.48	0.50	0.33

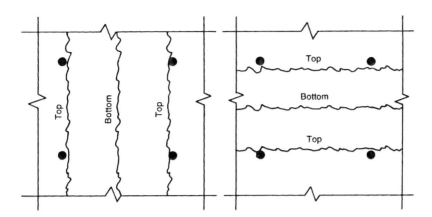

Fig. 3.24. Possible failure modes of flat slabs

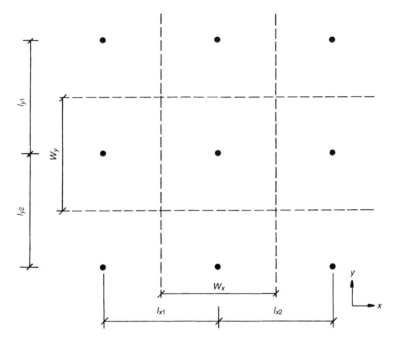

Fig. 3.25. Flat slabs: division into frames

When $l_x < l_y < 2l_x$:

$$W_x = (l_{x1} + l_{x2})/2$$

$$W_y = (l_{y1} + l_{y2})/2$$

When $l_y > 2l_x$:

$$W_x = (l_{x1} + l_{x2})/2$$

$$W_y = 2W_x$$

Horizontal loading. Horizontal loading in the frame will be considered only in unbraced structures. In these cases, the question of restraint to the columns, and hence the effective length of columns, is a matter of judgement. If the stiffness of the slab framing into the column is overestimated, the effective length of the column will reduce correspondingly. As the stiffness at slab–column junctions is a grey area, codes of practice adopt a cautious approach. For the slab, half the stiffness applicable to vertical loading is used.

Stiffness properties are generally based on the gross cross-section (ignoring the reinforcement). Additional stiffening effects of drops or solid concrete around columns in coffered slabs may be included, but this will complicate hand calculations. It should be noted that drops (and solid areas) should only be taken into account when the smaller dimension of the drop (and solid areas) is at least 33% of the smaller dimension of the surrounding panels.

Analysis
The equivalent frames may be analysed using any of the standard linear elastic methods such as moment distribution (see Section 3.7.1). Braced structures may be partitioned into subframes consisting of the slab at one level continuous with columns above and below. The far ends of the columns are normally taken as fixed unless this assumption is obviously wrong (e.g. columns with small pad footings not designed to take moments).

The load combinations given in Section 3.2 may be used.

The bending moments obtained from the analysis should be distributed laterally in the 'width' of the slab in accordance with the allocation of moments between strips (see p. 47).

Simplified coefficients
In braced buildings with at least three approximately equal bays and both slabs subject predominantly to uniformly distributed loads, the bending moments and shear forces may be obtained using the coefficients given in Section 3.8.1 (Table 3.3).

The bending moments obtained from the above, should be distributed laterally in the 'width' of the slab in accordance with h the allocation of moments between strips (see p. 47, Table 3.7).

Lateral distribution of moments in the width of the slab
In order to control the cracking of the slabs under service conditions, the bending moments obtained from the analysis should be distributed taking into account the elastic behaviours of the slab. As can be imagined, the strips of the slab on the lines of the columns will be stiffer than those away from the columns. Thus, the strips closer to the column lines will attract higher bending moments.

Division of panels
Flat slab panels should be divided into column and middle strips, as shown in Fig. 3.26.

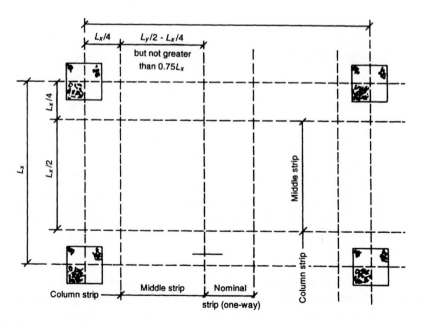

Fig. 3.26. Flat slabs: definition of panels

Table 3.7. Distribution of design moments in panels of flat slabs

	Apportionment between column and middle strip expressed as a percentage of the total negative or positive design moment	
	Column strip (%)	Middle strip (%)
Negative	60–80	40–20
Positive	50–70	50–30

For the case where the width of the column strip is taken as equal to that of the drop, and the middle strip is thereby increased in width, the design moments to be resisted by the middle strip should be increased in proportion to its increased width. The design moments to be resisted by the column strip may be decreased by an amount such that the total positive and the total negative design moments resisted by the column strip and middle strip together are unchanged.

As drawn, Fig. 3.26 applies to slabs without drops (or solid areas around columns in coffered slabs). When drops (or solid areas around columns in coffered slabs) of plan dimensions greater than $l_x/3$ are used, the width of the column strip may be taken as the width of the drop. The width of the middle strip should be adjusted accordingly.

Allocation of moments between strips
The bending moments obtained from the analysis should be distributed between the column and middle strips in the proportions shown in Table 3.7.

In some instances the analysis may show that hogging moments occur in the centre of a span (e.g. the middle span of a three-bay structure, particularly when it is shorter than the adjacent spans). The hogging moment may be assumed to be uniformly distributed across the slab, if the negative moment at midspan is less than 20% of the negative moment at supports. When the above condition is not met, the moment is concentrated more in the middle strip.

Moment transfer at edge columns
As a result of flexural and torsional cracking of the edge (and corner) columns, the effective width through which moments can be transferred between the slabs and the columns will be much narrower than in the case for internal columns. Empirically, this is allowed for in design by limiting the maximum moment the slab (without edge beams) can transfer to the columns:

$$M_{t\,max} = 0.17 b_e d^2 f_{ck}$$

where b_e is the effective width of the strip transferring the moment as defined in Fig. 3.27, and d is the effective depth of the slab.

$M_{t\,max}$ should not be less than 50% of the design moment obtained from an elastic analysis or 70% of the design moment obtained from grillage or finite element analysis. If $M_{t\,max}$ is less than these limits, the structure should be redesigned.

When the bending moment at the outer support obtained from the analysis exceeds $M_{t\,max}$, then the moment at the outer support should be reduced to $M_{t\,max}$, and the span moment should be increased accordingly.

The reinforcement required in the slab to transfer the outer support moment to the column should be placed on a width of slab $C_x + 2r$ (Fig. 3.28). For slabs with a thickness less than 300 mm, $r = C_y$, and for thicker slabs, $r = 1.67 C_y$. In the latter case, it is essential to provide torsional links along the edge of the slab. However, U bars (as distinct from L bars) with longitudinal anchor bars in the top and bottom may be assumed to provide the necessary torsional reinforcement (Fig. 3.29).

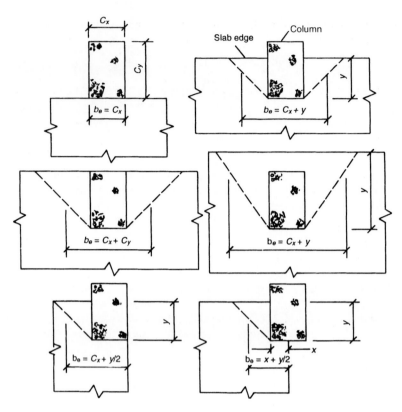

Fig. 3.27. Effective slab widths for moment transfer

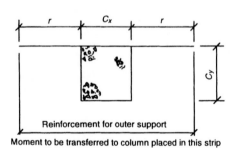

Fig. 3.28. Flat slabs: detailing at outer supports

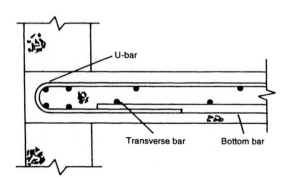

Fig. 3.29. Flat slabs: detailing at slab edges

Bending moments in excess of $M_{t\,max}$ may be transferred to the column only if an edge beam (which may be a strip of slab) is suitably designed to resist the tension.

3.8.4. Beams
For the design of beams, the bending moment coefficients given in Fig. 3.20 or Table 3.3 may be used, noting the conditions to be complied with, which are discussed in Section 3.8.1.

3.8.5. Simplifications
EN 1992-1-1 permits the following simplifications regardless of the method of analysis used:

(1) At a support assumed to offer no restraint to rotation (e.g. over walls), a beam or a slab which is continuous over it may be designed for a support moment which is less than the moment theoretically calculated on the centre line of the support. The permitted reduction in moment is then $F_{Ed,\,supp}t/8$, where $F_{Ed,\,supp}$ is the design support reaction and t is the breadth of the support. This recognizes the effect of the width of support and arbitrarily rounds off the peak in the bending moment diagram.

(2) Where a beam or a slab is cast monolithically into its supports, the critical moment may be taken as that at the face of the supports (but see also (3) below). This provision is quite reasonable, as failure cannot occur within the support.

(3) The design moment at the faces of rigid supports should not be less than 65% of the support moment calculated assuming full fixity at the faces of support. This ensures a minimum design value for the support moment, particularly in the case of wide supports.

(4) Loads on members supporting one-way-spanning continuous slabs (solid and ribbed) and beams (including T beams) may be assessed on the assumption that supports offer no rotational restraint. This is reasonable; but the effect of continuity should be considered when designing the support such as columns or walls.

CHAPTER 4

Materials and design data

4.1. Concrete

4.1.1. General

EN 1992-1-1 covers concrete of strength up to 90 MPa in normal-weight concrete and 80 MPa in lightweight concrete. It relies on EN 206-1 for the specification and production of concrete.

4.1.2. Strength

In EN 1992-1-1 the compressive strength of concrete is denoted by concrete strength classes, which relate to the characteristic cylinder strength f_{ck} or the cube strength $f_{ck, cube}$. The characteristic strength is defined as that strength below which not more than 5% of all test results are likely to fall. The relationship between the cylinder and cube strengths is given in the code, and is reproduced in Table 4.1.

The code provides models for strength development with time, and these should be used when concrete strength need to be calculated.

Table 4.1 also gives the mean tensile strength of concrete.

4.1.3. Elastic deformation

EN 1992-1-1 recognizes that deformation properties are crucially dependent on the composition of concrete, and in particular on the aggregates. Therefore, it provides only indicative information, which should be sufficient for most normal structures. However, in structures that are likely to be sensitive to deformation, it is advisable to determine the properties by controlled testing and appropriate specification.

Table 4.1 reproduces the mean values for the modulus of elasticity given in the code. The values given apply to concretes with quartzite aggregates. The reduction factors for other

Table 4.1. Strength classes and associated properties

	$f_{ck, cy}$ (MPa)												
	15	20	25	30	35	40	45	50	55	60	70	80	90
$f_{ck, cube}$ (MPa)	20	25	30	37	45	50	55	60	67	75	85	95	105
f_{cm} (MPa)	24	28	33	38	43	48	53	58	63	68	78	88	98
f_{ctm} (MPa)	1.9	2.2	2.6	2.9	3.2	3.5	3.8	4.1	4.2	4.4	4.6	4.8	5.0
$f_{ctk, 0.05}$ (MPa)	1.3	1.5	1.8	2.0	2.2	2.5	2.7	2.9	3.0	3.1	3.2	3.4	3.5
$f_{ctk, 0.95}$ (MPa)	2.5	2.9	3.3	3.8	4.2	4.6	4.9	5.3	5.5	5.7	6.0	6.3	6.6
E_{cm} (GPa)	29	30	31	33	34	35	36	37	38	39	41	42	44
ε_{cu2} (‰)	3.5	3.5	3.5	3.5	3.5	3.5	3.5	3.5	3.1	2.9	2.7	2.6	2.6

aggregates are as follows: limestone aggregates, 10%; basalt aggregates, 20%; and sandstone aggregates, 30%.

Poisson's ratio may be taken as 0.2 for uncracked concrete, and 0 for cracked concrete. The coefficient of linear thermal expansion may be taken as $10 \times 10^{-6}/K$.

4.1.4. Creep and shrinkage

The creep and shrinkage of concrete depend on a number of factors, including the ambient humidity, the dimensions of the element, the composition of the concrete and the age of concrete at the time of loading. The long-term values of the creep coefficient $\varphi(\infty, t_0)$ are shown in Figs 4.1 and 4.2 for inside and outside conditions, respectively. The resulting strain in the member will be $\varphi(\infty, t_0)(\sigma_c/E_c)$. When the compressive stress at time t_0 exceeds $0.45f_{ck}(t_0)$, the creep coefficient should be multiplied by a factor shown in Table 4.2, depending on the ratio

$$k_\sigma = \sigma_c/f_{cm}(t_0)$$

to allow for the non-linearity of creep development.

Shrinkage has two components, namely drying shrinkage (ε_{cd}) and autogenous shrinkage (ε_{ca}). Drying shrinkage is the result of expiration of moisture from the concrete to the surrounding air. On the other hand, the background document to EN 1992-1-1 defines autogenous shrinkage as 'the macroscopic volume reduction of cementitious materials when cement hydrates, after initial setting. It does not include the volume change due to loss or ingress of substances, temperature variation or the application of external force and restraint'. Autogenous shrinkage is related inversely to the water/cement ratio, and will be significant for concretes of high strength.

The drying shrinkage strain at age t is $\beta_{ds}(t, t_s)k_h\varepsilon_{cd, 0}$. Tables 4.3, 4.4 and 4.5 give the data for calculating shrinkage at 30, 100 and 1000 days, respectively.

Autogenous shrinkage strain at age t is given by $\varepsilon_{ca}(t) = \beta_{as}\varepsilon_{ca}(\infty)$. Values at 30, 100 and 1000 days are given in Table 4.6.

4.1.5. Stress–strain relationships

A distinction should be made between the stress–strain curve used for analysis and that used for the design of cross-sections. Figure 4.3 defines the parameters for the curve used for analysis. For numerical values, see EN 1992-1-1.

EN 1992-1-1 offers two alternatives for the design of cross-sections. These are parabola–rectangle (Fig. 4.4) and bi-linear (Fig. 4.5) diagrams. For numerical values of the parameters, see EN 1992-1-1.

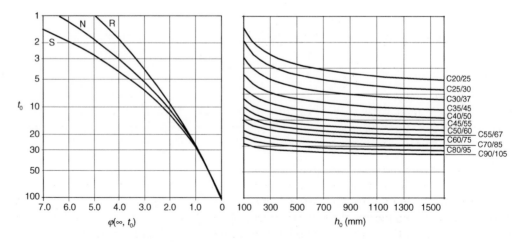

Fig. 4.1. Creep coefficients for concrete in indoor conditions

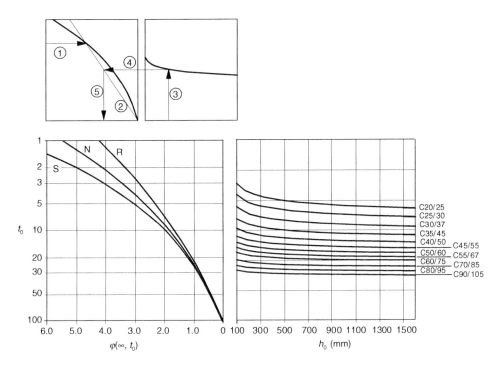

Fig. 4.2. Creep coefficients for concrete in outdoor conditions

Table 4.2. Multipliers for non-linear creep

k_σ	Multiplier
0.5	1.078
0.6	1.252
0.7	1.455
0.8	1.690
0.9	1.964
1.0	2.282

Table 4.3. Values of ε_{cd} (‰) for drying shrinkage

$f_{ck, cy}/f_{ck, cube}$ (MPa)	Relative humidity	
	Indoor conditions, 50%	Outdoor conditions, 80%
20/25	0.55	0.31
40/50	0.44	0.25
60/75	0.35	0.20
80/95	0.29	0.16
90/105	0.26	0.15

EN 1992-1-1 also permits a rectangular stress block to be used for section design. This is shown in Fig. 4.6. The basic compressive stress is $f_{cd} = \alpha_{cc} f_{ck}/\gamma_c$. The coefficient α_{cc} is said to take account of long-term effects on the compressive strength and of the unfavourable effects resulting from the way the load is applied. EN 1992-1-1 recommends a value of 1.0 for this coefficient, but the UK in its National Annex proposes 0.85. This reduced value, which was used in the ENV version of EN 1992-1-1, is considered necessary as a calibration factor

Table 4.4. Multipliers k_h to shrinkage coefficient for size of members

h_0 (mm)	k_h
100	1.00
200	0.85
300	0.75
≥ 500	0.70

(1) h_0 is the notional size of the cross-section in millimetres.
(2) $h_0 = 2 \times$ (area of cross-section)/(perimeter of that part of the cross-section which is exposed to drying).

Table 4.5. Multipliers $\beta_{ds}(t, t_s)$ for shrinkage development

h_0 (mm)	$t - t_s$ days		
	30 days	100 days	1000 days
100	0.428	0.714	0.962
200	0.209	0.469	0.898
300	0.126	0.325	0.828
≥ 500	0.063	0.183	0.691

$t - t_s$ is the age of concrete in days from the end of curing when the shrinkage strain is being calculated.

Table 4.6. Autogenous shrinkage coefficients at different ages $\varepsilon_{ca}(\infty)\beta_{as}(t)$ (‰)

	f_{ck} (MPa)										
	20	25	30	35	40	45	50	60	70	80	90
$\varepsilon_{ca}(\infty)\beta_{as}$ (1000 days)	0.025	0.038	0.050	0.063	0.075	0.088	0.100	0.125	0.150	0.175	0.200
$\varepsilon_{ca}(\infty)\beta_{as}$ (100 days)	0.022	0.033	0.043	0.054	0.065	0.075	0.086	0.108	0.129	0.151	0.172
$\varepsilon_{ca}(\infty)\beta_{as}$ (30 days)	0.017	0.025	0.033	0.042	0.050	0.056	0.067	0.083	0.100	0.116	0.133

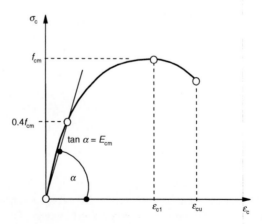

Fig. 4.3. Stress–strain relation for structural analysis

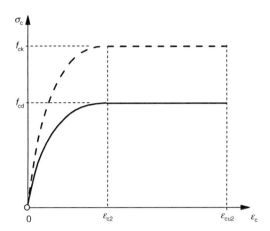

Fig. 4.4. Parabola–rectangle diagram for concrete under compression

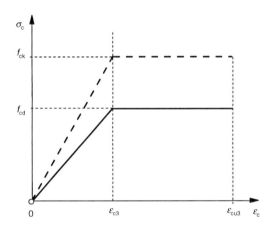

Fig. 4.5. Bi-linear stress–strain relation for concrete

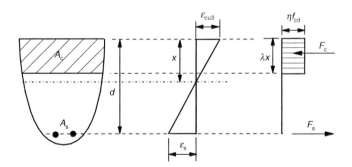

Fig. 4.6. Rectangular stress distribution for concrete

between the predicted strength and that obtained in experiments. See the background paper to the UK National Annex.[6]

The other parameters defining the rectangular stress block are λ and η. These are functions of concrete strength, but have constant values of 0.8 and 1.0, respectively, for $f_{ck} \leq 50$ MPa. Table 4.7 lists the values for different concrete grades.

The value of design tensile strength $f_{ctd} = \alpha_{ct} f_{ctk, 0.05}/\gamma_c$. In the UK National Annex, $\alpha_{ct} = 1.00$.

4.2. Lightweight concrete

4.2.1. General

EN 1992-1-1 deals with additional requirements for lightweight concrete in Chapter 11. The requirements for normal-weight concrete are generally applicable to lightweight concrete unless specifically varied. The code uses the subscript 'l' to distinguish lightweight concrete (f_{lck}, E_{lcm}, ε_{lcu2}, etc.)

Lightweight concrete is defined as concrete having a closed structure and a density not exceeding 2200 kg/m^3.

4.2.2. Density classes

Six density classes are defined in EN 206-1. Table 4.8 reproduces the relevant data. In each class, a range is given for the density and the nominal density to be used in design calculations.

Many properties of lightweight concrete are related to its density ρ. The coefficient

$$\eta_1 = 0.4 + 0.6\rho/2200$$

is used to modify the relevant property of normal-weight concrete. In this expression, ρ refers to upper limit of the density for the relevant density class.

Table 4.7. Values of λ and η for different strengths of concrete

f_{ck} (MPa)	λ	η
≤ 50	0.800	1.00
60	0.775	0.95
70	0.750	0.90
80	0.725	0.85
90	0.700	0.80

Table 4.8. Density classes for lightweight concrete

	Density class					
	1.0	1.2	1.4	1.6	1.8	2.0
Density range (kg/m^3)	801–1000	1001–1200	1201–1400	1401–1600	1601–1800	1801–2000
Nominal design density (kg/m^3)						
Plain concrete	1050	1250	1450	1650	1850	2050
Reinforced concrete	1150	1350	1550	1750	1950	2150

Table 4.9. Strength classes and associated properties for lightweight concrete

	$f_{lck, cy}$ (MPa)											
	16	20	25	30	35	40	45	50	55	60	70	80
$f_{lck, cube}$ (MPa)	18	22	28	33	38	44	50	55	60	66	77	88
ht f_{lcm} (MPa)	22	28	33	38	43	48	53	58	63	68	78	88
f_{lctm} (MPa)	$f_{lctm} = f_{ctm}\eta_1$											
$f_{lctk, 0.05}$ (MPa)	$f_{lctk, 0.05} = f_{ctk, 0.05}\eta_1$											
$f_{lctk, 0.95}$ (MPa)	$f_{lctk, 0.95} = f_{ctk, 0.95}\eta_1$											
E_{lcm} (GPa)	$E_{lcm} = E_{cm}\eta_E$											
ε_{lcu2} (‰)	$3.5\eta_1$	$3.5\eta_1$	$3.5\eta_1$	$3.5\eta_1$	$3.5\eta_1$	$3.5\eta_1$	$3.5\eta_1$	$3.5\eta_1$	$3.1\eta_1$	$2.9\eta_1$	$2.7\eta_1$	$2.6\eta_1$

See the text for the definition of η_1 and η_E.

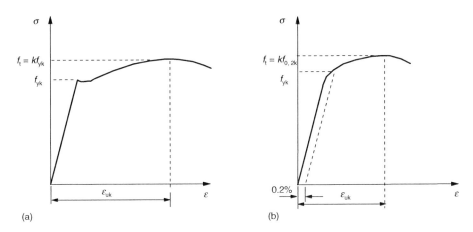

Fig. 4.7. Stress–strain diagrams of typical reinforcing steel. (a) Hot-rolled steel. (b) Cold-worked steel

4.2.3. Other modifying coefficients
- The mean value of the modulus of elasticity is obtained by multiplying the value for normal weight concrete by $\eta_{E} = (\rho/2200)^{2}$.
- The creep coefficient $\varphi(\infty, t_{0})$ for normal-weight concrete should also be multiplied by η_{E} to obtain the coefficient for lightweight concrete. The creep strains so derived should be further multiplied by 1.3 for $f_{lck} \leq LC16/20$.
- Shrinkage strains for lightweight concrete should be obtained by multiplying the values of $\varepsilon_{cd, 0}$ for normal-weight concrete by $\eta_{3} = 1.5$ for $f_{lck} \leq LC16/20$ and 1.2 for $f_{lck} \geq LC20/25$.

Table 4.9 summarizes the properties of lightweight concrete.

4.2.4. Design compressive strength
The value of the design compressive strength is

$$f_{lcd} = \alpha_{lcc} f_{lck}/\gamma_{c}$$

The recommended value in the code for α_{lcc} is 0.85. The UK National Annex proposes to adopt the same value.

Similarly, the value of the design tensile strength is

$$f_{lctd} = \alpha_{lct} f_{lctk}/\gamma_{c}$$

with $\alpha_{lct} = 0.85$.

4.3. Reinforcing steel
EN 1992-1-1 rules are valid for ribbed bars, de-coiled rods, welded fabrics and lattice girders. The range of essential properties is listed in the normative *Annex C* of the code. Although the reinforcement standard EN 10080 will be published soon, it will confine itself to the testing requirements of properties and not define the properties themselves.

4.3.1. Strength (f_{yk})
The grade of reinforcement steel denotes the specified characteristic yield stress (f_{yk}). It is obtained by dividing the characteristic yield load by the nominal cross-section area of the bar. For products without a pronounced yield stress, the 0.2% proof stress $f_{0.2k}$ is substituted as the yield stress. Typical and idealized stress–strain diagrams are given in Fig. 4.7. EN 1992-1-1 is valid for yield stress of reinforcement in the range 400–600 MPa. In the UK, the characteristic strength (f_{yk}) of the commonly used grade of reinforcement is likely to be 500 MPa.

In general, ductility is inversely related to yield stress. Therefore, in applications where ductility is critical (e.g. seismic design), it is important to ensure that the actual yield strength

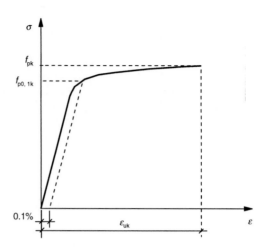

Fig. 4.8. Stress–strain diagram for typical prestressing steel

Table 4.10. Ductility classes for bars, decoiled rods and wire fabrics

	Ductility class A	Ductility class B	Ductility class C
$k = (f_t/f_y)_k$	≥ 1.05	≥ 1.08	≥ 1.15 but < 1.35
ε_{uk} (%)	≥ 2.5	≥ 5.0	≥ 7.5

does not exceed the specified value by a large margin. Limits for the ratio of the actual yield stress to the specified strength are specified for such applications. Currently, no UK standard gives any guidance.

There is no technical reason why other types of reinforcement should not be used in conjunction with EN 1992-1-1, provided suitable allowance is made for the behaviour. For instance, the UK National Annex provides some advice on the use of plain bars with $f_{yk} = 250$ MPa. Relevant authoritative publications should be consulted when other types of reinforcement are used.

4.3.2. Ductility
Ductility is defined using the strain at maximum load (ε_{uk}) and the ratio between the maximum and the yield strengths $(f_t/f_y)_k$. Ductility is an essential property if advantage is to be taken of the plastic behaviour of structures. The greater the ductility, the greater the elongation in axially loaded members, and the greater the rotation capacity in members subjected to flexure. In members where the ultimate strength is governed by the yielding of reinforcement (shallow members with a low percentage of steel), clearly the higher the value of ε_{uk} the greater the ductility (i.e. the plateau of the stress–strain diagram is long).

When the ultimate strength is controlled by the strain in concrete reaching the limiting value, the length of the plastic zone influences the rotation. The longer this length, the greater is the rotation. f_t/f_y indicates the length over which yield takes place (i.e. the length of the plastic zone) (Fig. 4.8).

EN 1992-1-1 defines three ductility classes (A, B and C (in terms of ε_{uk} and (f_t/f_y)). These are shown in Table 4.10.

4.4. Prestressing steel
EN 1992-1-1 states that its requirements will be satisfied if the prestressing steel complies with EN 10138. The definition of the relevant parameters is shown in Fig. 4.8.

CHAPTER 5

Design of sections for bending and axial force

This chapter is concerned with the design for the ultimate limit state of sections subject to pure flexure, such as beams or slabs, sections subject to combined bending and axial load, such as columns, and prestressed sections. The material in this chapter is covered in EN 1992-1-1 in the following clauses:

- Design stress–strain curves for reinforcement   *Clause 3.2.3*
- Design stress–strain curve for prestressing steel   *Clause 3.3.3*
- Design stress–strain curves for concrete   *Clause 3.1.7*
- Basic assumptions for section design   *Clause 6.1*
- Minimum reinforcement   *Clauses 7.3.2* and *9.2.1.1*
- Limitations imposed by redistribution   *Clause 5.5*

5.1. Basic assumptions

The basic assumptions about section behaviour are very similar to those adopted by many, if not most, modern codes of practice. The formulation used in EN 1992-1-1 is taken from the *CEB Model Code for Concrete Structures*.[7] The assumptions in EN 1992-1-1 differ in detail to those in the UK code, BS 8110 in ways which, as will be seen, make calculations rather more complex in some cases, but the practical outcome is not significantly different. The assumptions define the stress–strain responses to be assumed for steel and concrete, and the assumptions to be made about the strains at the ultimate limit state. It is these assumptions about the strain that define failure.

5.1.1. Stress–strain curves

The information required to obtain the design stress–strain curves for concrete, ordinary reinforcement and prestressing steel are to be found in *clauses 3.1.7*, *3.2.3* and *3.3.3*, respectively, of EN 1992-1-1.

Clause 3.1.7
Clause 3.2.3
Clause 3.3.3

For reinforcement and prestressing steel the code specifies the use of bi-linear stress–strain curves. These are given in *clauses 3.2.3* and *3.3.3* for reinforcing and prestressing steels, respectively. In each case, it is possible to choose between two possible bi-linear diagrams for the design of sections: one with a horizontal top branch and one with an inclined top branch. These curves are shown in Fig. 5.1. Where the horizontal top branch is used, no limit on the tensile strain is imposed; however, the characteristics of the inclined top branch depend on the ductility class of the reinforcement. These are given in *Table 3.3* in EN 1992-1-1 for reinforcement, and are repeated in Table 5.1 for convenience.

Clause 3.2.3
Clause 3.3.3

The permitted ultimate strain in design when the inclined upper branch is used is $0.9\varepsilon_{uk}$. This is assumed to correspond to a maximum design stress of $(f_t/f_y)_k f_{yk}/\gamma_s$. Clearly, use of the curves with inclined top branches will give some economic advantage over the use of the horizontal top branch. Potentially, this advantage could reach an 8% saving in reinforcement, but a saving approaching this value will only rarely be achievable, and will be at the expense of considerably more complex calculations. The full saving will only be available where it is certain at the design stage that a high-ductility steel will be used and where the neutral axis depth at the ultimate limit state will be around 0.25. Above this the steel stress will be below f_{tk}/γ_s, while for lower neutral axis depths the effect of limiting the tension strain will lead to a lower average stress in the compression zone, and hence a deeper neutral axis than if the strain was unlimited.

In order to keep the design equations and design aids reasonably simple, the horizontal upper branch will be used in the derivation of all equations, charts and tables in this chapter.

Clause 3.2.3

For concrete, three possibilities are described in *clause 3.2.3* of EN 1992-1-1. The preferred idealization is the parabolic–rectangular diagram, but a bi-linear diagram and a rectangular diagram are also permitted. These three diagrams are compared in Fig. 5.2 with the limiting strains tabulated in Table 5.2.

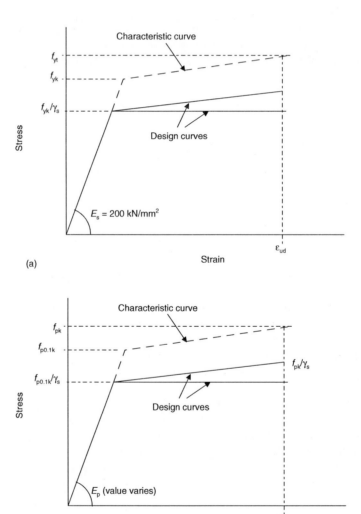

Fig. 5.1. (a) Stress–strain diagram for reinforcing steel. (b) Design and characteristic stress–strain curves for prestressing steel

Table 5.1. Parameters defining the ductility classes of reinforcement

Ductility class	Ultimate tensile strain, ε_{uk}	Ratio $(f_t/f_y)_k$
A	0.025	1.05
B	0.050	1.08
C	0.075	$1.15 \leq (f_t/f_y)_k \leq 1.35$

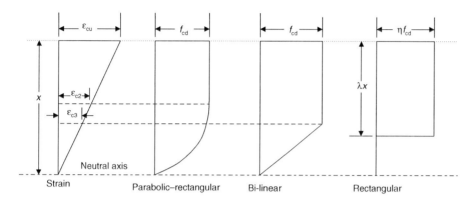

Fig. 5.2. Idealized stress-strain distributions

Table 5.2. Parameters defining design stress–strain curves for concrete

Grade	ε_{cu}	ε_{c2}	ε_{c3}	n	λ	η
$\leq$ C50	0.0035	0.0020	0.00175	2.0	0.800	1.000
C55	0.0031	0.0022	0.00180	1.75	0.788	0.975
C60	0.0029	0.0023	0.00190	1.60	0.775	0.950
C70	0.0027	0.0024	0.00200	1.45	0.750	0.900
C80	0.0026	0.0025	0.00220	1.40	0.725	0.850
C90	0.0026	0.0026	0.00230	1.40	0.700	0.800

Extension of the code to include concrete grades up to 90 N/mm² has required modification of the idealized stress–strain curves for concrete strengths greater than 50 N/mm². This arises because research has shown that higher concrete strengths exhibit more brittle behaviour. Above a characteristic strength of 50 N/mm², the parabolic–rectangular diagram is not, in fact, parabolic–rectangular even though the code describes it as such. The formula for the curved part of the response is

$$\sigma_c = f_{cd}[1 - (1 - \varepsilon_c/\varepsilon_{c2})^n]$$

n varies from 2 to 1.4, depending on the concrete strength, and is clearly only a parabola when $n = 2$.

It will be noted that the maximum design stress is given by the characteristic concrete strength multiplied by a coefficient α_{cc} and divided by the partial safety factor for the concrete, i.e.

$$f_{cd} = \alpha_{cc}f_{ck}/\gamma_c$$

The coefficient α_{cc} is described in *clause 3.1.6* in EN 1992-1-1 as '*taking account of long term effects on the compressive strength and of unfavourable effects resulting from the way the load is applied*'. A value of 1.0 is suggested, based on the argument that there should be no significant long-term effects at ultimate not already included in the data from which the stress–strain curves were derived. It is, however, arguable that the reason for the

Clause 3.1.6

introduction of this factor is more related to the idealization of the shape of the stress–strain diagram as used for flexure than to the nature of the loading (i.e. it is needed to allow for the way the load is applied). Figure 5.3 tries to illustrate this. It will be seen that, for the area under the actual curve and the idealized curve to be the same, the maximum stress level for the idealized curve must be below the maximum stress of the 'true' diagram. Furthermore, study of the data available on the behaviour of compression zones at failure suggests that the use of 1.0 is unconservative. For this reason, the UK National Annex recommends a value for α_{cc} of 0.85, as is proposed in the CEB Model Codes.

Derivation of stress block parameters

Table 5.3 compares the three permitted idealizations in terms of the average stress over a rectangular compression zone and the distance from the compression face of the section to the centre of compression as a fraction of the neutral axis depth (β). In this table, α has been assumed to be 0.85 and γ_c has been taken as 1.5. As well as providing a convenient comparison of the idealizations, Table 5.3 will be found to provide information in a useful form for design calculations.

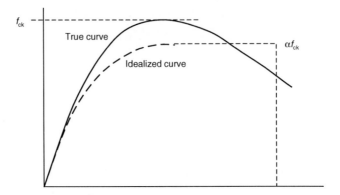

Fig. 5.3. Comparison of true and idealized stress–strain curves

Table 5.3. Comparison of stress block parameters

| Grade | Parabolic–rectangular | | Bi-linear rectangular | | Rectangular | |
	Average stress (N/mm²)	Centroid factor, β	Average stress (N/mm²)	Centroid factor, β	Average stress (N/mm²)	Centroid factor, β
12	5.51	0.416	5.10	0.389	5.44	0.40
16	7.34	0.416	6.80	0.389	7.25	0.40
20	9.18	0.416	8.50	0.389	9.07	0.40
25	11.47	0.416	10.63	0.389	11.33	0.40
30	13.76	0.416	12.75	0.389	13.60	0.40
35	16.06	0.416	14.88	0.389	15.87	0.40
40	18.35	0.416	17.00	0.389	18.13	0.40
45	20.64	0.416	19.13	0.389	20.40	0.40
50	22.94	0.416	21.24	0.389	22.67	0.40
55	23.19	0.393	22.10	0.374	23.93	0.39
60	23.58	0.377	22.88	0.363	25.03	0.39
70	24.86	0.360	24.55	0.349	26.78	0.38
80	27.1	0.355	26.51	0.342	27.94	0.36
90	29.75	0.353	28.44	0.337	28.56	0.35

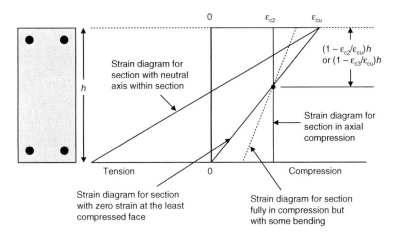

Fig. 5.4. Ultimate strain distributions

It will be seen from Table 5.3 that the results obtained from the three idealizations will be so similar as to be indistinguishable for all normal purposes for concrete strengths up to 50 N/mm^2.

5.1.2. Assumptions relating to the strains at the ultimate limit state

The basic assumptions here are that plane sections remain plane and that the strain in reinforcement is the same as the strain in the concrete at the same level. These assumptions are universally accepted for the design of members containing bonded ordinary reinforcement. For prestressed sections, allowance has to be made for the strain in the steel prior to it being bonded to the concrete. After bonding, the change in strain in the steel is assumed to be the same as the change in strain in the concrete. The assumption is, of course, invalid for beams with unbonded tendons. These lie outside the scope of EN 1992-1-1.

However, the assumptions are not strictly true. The deformations within a section are very complex, and, locally, plane sections do not remain plane. Nor, due to local bond slip, are the strains in the concrete exactly the same as those in the steel. Nevertheless, on average, the assumptions are correct, and are certainly sufficiently true for practical purposes for the design of normal members. One area where they are not adequate is in short members. For this reason, deep beams are not designed using the provisions of this chapter, and nor are members such as corbels.

5.2. Limiting compressive strains

It is universal to define failure of concrete in compression by means of a limiting compressive strain. The formulation of the limit varies from code to code, for example the American Concrete Institute code, ACI 318, uses a limit of 0.003, while the UK code BS 8110 uses 0.0035. For concrete strengths not exceeding 50 N/mm^2, the Eurocode adopts values taken from the CEB Model Code. These comprise a limit of 0.0035 for flexure and for combined bending and axial load where the neutral axis remains within the section, and a limit of between 0.0035 and 0.002 for sections loaded so that the whole section is in compression. This is illustrated in Fig. 5.4. The values for the ultimate compressive strains for situations where the neutral axis lies within the section and for axial compression vary for higher-strength concretes, with the ultimate strain reducing with increasing strength for cases where the neutral axis is within the section, while the value for axial compression increases with increasing strength.

The logic behind the reduction in the strain limit for axial compression is that, in axial compression, failure will occur at the strain corresponding to the attainment of the maximum compressive stress. This is 0.002 for concrete strengths not exceeding 50 N/mm². In flexure, considerably higher strains can be reached before the maximum capacity of the section is reached, and the value of 0.0035 has been obtained empirically. A means is needed to interpolate between the value of 0.0035 for flexure and 0.002 for axial load, and Fig. 5.4 provides this.

5.2.1. Singly reinforced beams and slabs

The conditions in a singly reinforced rectangular section at the ultimate limit state are assumed to be as shown in Fig. 5.5.

From Fig. 5.5, it will be seen that the following equations can be derived from consideration of equilibrium of axial forces and moments:

$$f_{av}bx = f_{yd}A_s \qquad \text{or} \qquad x/d = \rho f_{yd}/f_{av} \tag{D5.1}$$

$$M = f_{av}bx(d - \beta x) \qquad \text{or} \qquad M/bd^2 = f_{av}(1 - \beta x/d)x/d \tag{D5.2}$$

where A_s is the area of tension steel, f_{av} is the average stress over the compression zone, f_{yd} is the design yield strength of the steel, β is the ratio of the distance of the centre of compression from the compression face to the neutral axis depth, and ρ is the reinforcement ratio (A_s/bd).

For the parabolic–rectangular stress–strain curve for a concrete strength not greater than 50 N/mm² and taking the partial safety for concrete as 1.5 and α as 0.85, we can write

$$f_{av} = 0.459f_{ck}$$

$$f_{yd} = f_{yk}/1.5$$

$$\beta = 0.416$$

Values for higher strength concretes can be obtained from Table 5.3.

Substituting for x/d in equation (D5.2) from equation (D5.1) gives

$$\rho f_{yd}/f_{av} = 1/2\beta - \surd(1/4\beta^2 + K_{av}/\beta)$$

where $K_{av} = M_d/bd^2f_{av}$ or, substituting the values for β and f_{av} given above,

$$A_s = f_{ck}bd(0.633 - \surd(0.4 - 1.46K)/f_{yk}) \tag{D5.3}$$

in which $K = M/bd^2f_{ck}$.

It may also sometimes be useful to be able to obtain the neutral axis depth and the lever arm. These are given by

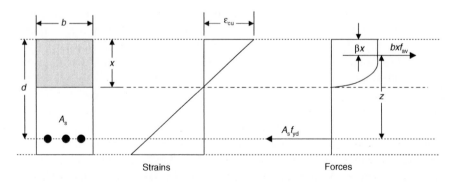

Fig. 5.5. Conditions in a singly reinforced section at the ultimate limit state

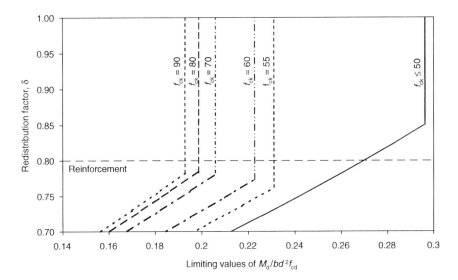

Fig. 5.6. Limiting values of M_d/bd^2f_{cd} as a function of the amount of redistribution

$$x/d = 1.895f_{yk}A_s/(f_{ck}bd) \tag{D5.4}$$

$$z/d = 1 - 0.416x/d \tag{D5.5}$$

It is now necessary to consider the limits to the use of singly reinforced sections. *Clause 5.5* in EN 1992-1-1 gives limits to the neutral axis depth at the ultimate limit state as a function of the amount of redistribution carried out in the analysis. The limit is repeated here for convenience:

$$x/d \le (\delta - 0.4)/(0.6 + 0.0014/\varepsilon_{cu}) \tag{D5.6}$$

In applying this formula, some degree of interpretation of the code is necessary since the reader is referred to *Table 3.1* of EN 1992-1-1 for values of the ultimate strain. *Table 3.1* gives several values, and it has been assumed that the ultimate strain is ε_{cu1} in this table. x/d should not be greater than 0.45 for concrete grades C50 or below, and not greater then 0.35 for grades C55 or higher.

δ in equation (D5.6) is the ratio of the redistributed moment to the moment before redistribution. δ is limited as a function of the type of reinforcement used as follows: for Class B and Class C steel, $\delta \ge 0.70$ and for Class A steel, $\delta \ge 0.8$.

The neutral axis limits set out above can now be substituted into equation (D5.2), to give limiting values of the parameterized moment, M_d/bd^2f_{cd}. This gives the limits shown in Fig. 5.6. The sloping parts of the relationships for a given concrete strength are the parts of the relationship controlled by the equation given above while the vertical parts are governed by the neutral axis limits of 0.45d or 0.35d as appropriate for the particular concrete strength.

The expressions for the calculation of the areas of reinforcement can be presented as design charts. This is done in Fig. 5.7, which gives the mechanical reinforcement ratio, Af_{yd}/bdf_{cd}, in terms of the non-dimensional moment parameter, M/bd^2f_{cd}. It should be noted that M/bd^2f_{cd} should not be taken as greater than the limiting value given by Fig. 5.6. The scale on the upper edge of Fig. 5.7 gives values of x/d at ultimate. This is of use where flanged sections are being considered. If a greater moment capacity is required, then a doubly reinforced section will be required. This is considered below. It can easily be shown that there is no need to distinguish between different concrete grades, as the line for a 90 N/mm² concrete is indistinguishable from that for a 50 N/mm² concrete. This chart can now be used for the design of singly reinforced rectangular sections or flanged sections where the neutral axis remains within the flange. The use of the chart is illustrated below by a simple example.

Example 5.1: singly reinforced rectangular section
A rectangular section 500 mm deep by 300 mm wide is to be designed to carry an ultimate moment of 120 kN m. The characteristic strength of the reinforcement is 500 N/mm², and that of the concrete is 25 N/mm². Fifteen per cent of the redistribution was carried out in the analysis for the ultimate limit state.

Assume that the effective depth is 450 mm. Assuming that the partial safety factor for the concrete and the reinforcement are, respectively, 1.5 and 1.15, and α_{cc} is 0.85, f_{cd} is then 14.17, and f_{yd} is 435 N/mm². This enables M/bd^2f_{cd} to be calculated as

$$120 \times 10^6/(300 \times 450^2 \times 14.17) = 0.139$$

From Fig. 5.7, the mechanical reinforcement ratio can be read off as 0.151, giving an area of reinforcement of

$$0.151 \times 300 \times 450 \times 14.17/435 = 663 \text{ mm}^2$$

A check should also be carried out to ensure that M/bd^2f_{cd} is less than the limiting value for the amount of redistribution carried out. Since 15% of redistribution has been done, $\delta = 0.85$, and hence, from Fig. 5.6, the maximum allowable value of M/bd^2f_{cd} is 0.296. This is greater than 0.139, and hence a singly reinforced section is satisfactory.

5.2.2. Doubly reinforced rectangular sections

If the design moment is greater than the limiting value obtained from Fig. 5.6, it is necessary to add compression reinforcement to maintain the neutral axis depth at the limiting value. The equations

$$A_{sc} = (M - M_{lim})/f_{yd}(d - d') \tag{D5.7}$$

$$A_s = A_{sc} + A_{s,lim} \tag{D5.8}$$

will achieve this. M_{lim} is the moment corresponding to the limiting value of M/bd^2f_{cd} obtained from Fig. 5.6. $A_{s,lim}$ is the value of A_s obtained when $M = M_{lim}$. This can be obtained from Fig. 5.7.

It should be noted that the equations derived in this section implicitly assume firstly that the area displaced by the compression reinforcement is ignored and secondly that the compression reinforcement is working at its design yield strength. This second assumption is only true if

$$f_{sd}/E_s < \varepsilon_{cu}(x - d')/x \tag{D5.9}$$

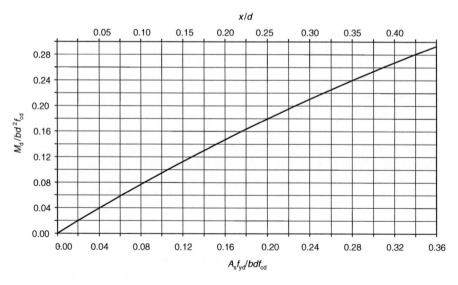

Fig. 5.7. Design chart for singly reinforced sections

Figure 5.7 in conjunction with equations (D5.7) and (D5.8) can be used for the design of doubly reinforced sections, as illustrated in Example 5.2.

Example 5.2: rectangular beam with compression reinforcement

The amounts of tension and compression reinforcement are to be calculated for a section 300 mm wide and 450 mm overall depth to sustain a design ultimate moment of 275 kN m. The distance from the compression face to the centroid of any compression steel and the distance from the tension face to the centroid of the tension steel may both be assumed to be 50 mm. The reinforcement has a characteristic strength of 500 N/mm^2, and the concrete is grade C30/37. Fifteen per cent of the redistribution was carried out in the analysis for the ultimate limit state.

f_{yd} and f_{cd} can be calculated as, respectively, 435 and 17 N/mm^2. The value of δ corresponding to 15% redistribution is 0.85. The effective depth is $450 - 50 = 400$ mm. Figure 5.6 can be used to obtain the limiting value of M/bd^2f_{cd} corresponding to this; it is 0.296, giving a value of M_{lim} of 241.5 kN m. Equation (D5.7) can now be used to obtain the area of compression steel: substitution into this equation gives

$$A_{sc} = (275 - 241.5) \times 10^6/[435 \times (400 - 50)] = 220 \text{ mm}^2$$

Figure 5.6 gives a value of $A_s f_{yd}/(bdf_{cd})$ corresponding to M_{lim} of 0.36. This corresponds to a reinforcement area of 1688 mm^2. Substitution into equation (D5.8) gives the area of tension reinforcement as

$$A_s = 220 + 1688 = 1908 \text{ mm}^2$$

5.2.3. Design of flanged sections (T or L beams)

Provided that the neutral axis at the ultimate limit state remains within the flange of a T or L beam, the section may be treated as a rectangular section, and the equations derived above may be used. From experience, it will only very infrequently be necessary to do other than consider a flanged beam as rectangular. If the neutral axis does need to be deeper than the flange depth, then it will be sufficiently accurate to assume that the whole of the outstanding parts of the flanges are carrying a concrete stress of f_{cd}. This leads to the equilibrium equations given below:

$$A_s f_{yd} = f_{cd} h_f(b - b_r) + f_{av} b_r x \tag{D5.10}$$

$$M = f_{av} b_r x(d - \beta x) + f_{cd} h_f(b - b_r)(d - h_f/2) \tag{D5.11}$$

where b is the overall breadth of the section, b_r is the breadth of the rib and h_f is the depth of the flange. An alternative way of looking at these equations is to see that the outstanding part of the flanges carry a moment equal to

$$M_{flange} = f_{cd} h_f(b - b_r)(d - h_f/2)$$

which requires a reinforcement area of

$$A_f = f_{cd} h_f(b - b_r)/f_{yd}$$

The remaining part of the beam can be considered as a rectangular beam of width b_r and effective depth d supporting a moment of $M - M_{flange}$. The reinforcement required for this may be calculated using Figs 5.6 and 5.7. The total area of tension reinforcement is then A_f plus that required to resist $M - M_{flange}$.

5.2.4. Checking the moment of resistance of more complex section shapes

It is not possible to derive equations for all possible situations, and it is useful to have a more general approach available. A suitable iterative approach for the calculation of the moment of resistance of any section is as follows:

(1) Guess a value for x.
(2) Calculate the compressive force in the compression zone using any of the allowable idealizations of the compressive stress–strain curve. If the parabolic–rectangular diagram is used, a numerical technique may be most convenient. For simple hand calculations, the rectangular idealization is much the easiest to apply.
(3) From the neutral axis depth and the ultimate strain, calculate the strain in each layer of reinforcement and hence the stress and the force in each layer.
(4) Check whether the total compressive force is equal to the total tension force. If this is so, then the neutral axis depth is correct, and the moments of the internal forces may be taken to obtain the moment of resistance. If the tension and compression forces are not equal, adjust the value of the neutral axis depth and repeat the calculation from step (2).

5.2.5. Design of rectangular column sections

The basic equations for equilibrium of a rectangular section subjected to combined bending and axial load are as given below for situations where the neutral axis remains within the section:

$$N_{Rd} = f_{av}bx + \sum f_s A_s \tag{D5.12}$$

$$M_{Rd} = f_{av}bx(h/2 - \beta x) + \sum f_s A_s(h/2 - d_i) \tag{D5.13}$$

In equation (D5.13), moments have been taken about the centroid of the concrete section. The summation signs indicate a summation over all the levels of reinforcement within the section. In carrying out the summation, tensile stresses must be taken as negative. d_i is the distance from the compressive face of the section to the ith layer of reinforcement.

Assuming that the values for the partial safety factors on the steel and concrete are taken as 1.15 and 1.5, respectively, and taking α_{cc} as 0.85, we can substitute $0.459f_{ck}$ for f_{av} and 0.416 for β. The resulting equations are rigorous for situations where the neutral axis remains within the section. Where the whole section is in compression, more complex expressions are required to deal with (1) the portion of the parabolic curve cut off by the bottom of the section and (2) the reduction in the ultimate strain at the compressive face according to Fig. 5.4. Additionally, more complex equations are required for cases where the concrete strength exceeds 50 N/mm². The resulting equations are unwieldy, and it is not appropriate to present them here. A simpler approach is to use design charts. Figures 5.8a-l provide a series of parameterized charts for the design of symmetrically reinforced rectangular columns derived using the parabolic–rectangular diagram. Three sets of charts are included here. The first set (Figs 5.8a–d) can be used for sections where either the reinforcement is concentrated at the corners, as shown in the diagrams on the charts, or for sections where the steel is distributed along the sides parallel to the axis of bending. The second set of charts (Figs 5.8e–h) are for sections where the reinforcement is distributed in the most disadvantageous way along the sides of the section which are perpendicular to the axis of bending. The final set of charts (Figs 5.8i–l) are for sections where the reinforcement is distributed evenly around the section perimeter. The use of these charts is illustrated by Example 5.3.

Rigorously, a further complete set of charts is required for each concrete grade above 50 N/mm², as the parameters defining the concrete stress–strain curve change. What has been done here is to provide a further complete set of charts for 90 N/mm² concrete (Figs 5.9a–l). Interpolation may be used between the charts for 50 and 90 N/mm² concrete for intermediate strengths.

Charts can also be produced for circular column sections, and a set of parameterized charts are given in Figs 5.10a–d. The charts are drawn assuming that the section contains six reinforcing bars, which is the minimum that can reasonably be used in a circular section. It is found that there is no single arrangement of the reinforcement relative to the axis of bending which will always give the minimum strength. The charts are therefore drawn to give a lower bound envelope to the interaction diagrams for various arrangements of the bars.

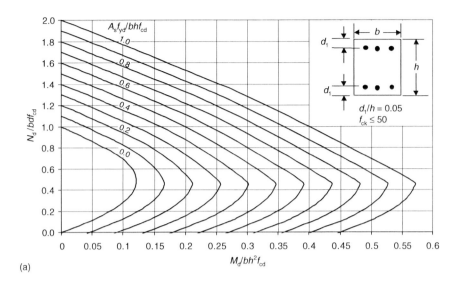

(a)

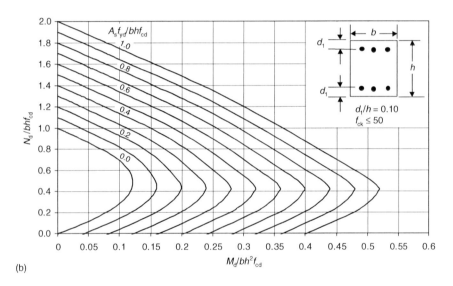

(b)

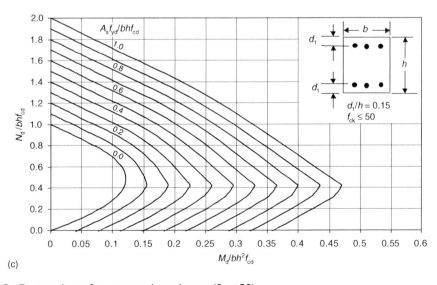

(c)

Fig. 5.8. Design charts for rectangular columns ($f_{ck} \leq 50$)

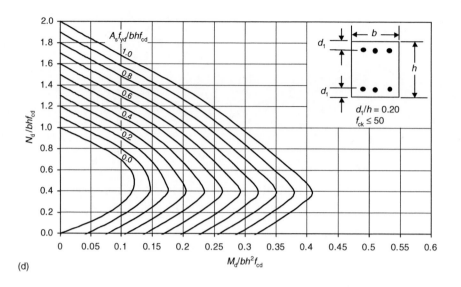

(d)

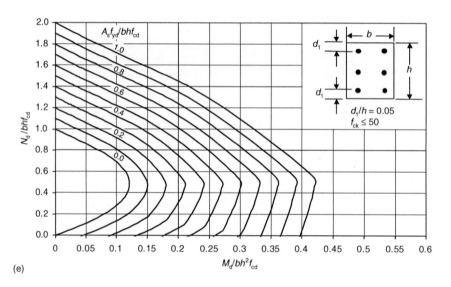

(e)

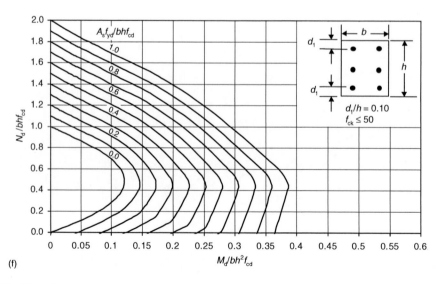

(f)

Fig. 5.8. (Contd)

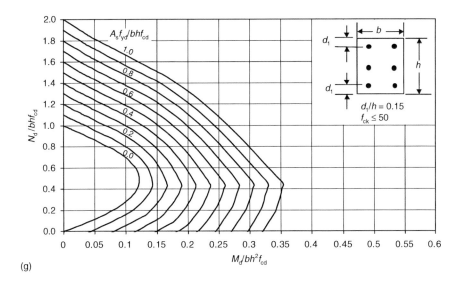

(g)

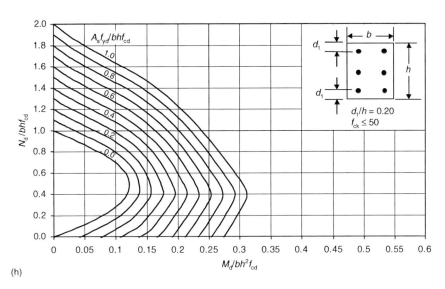

(h)

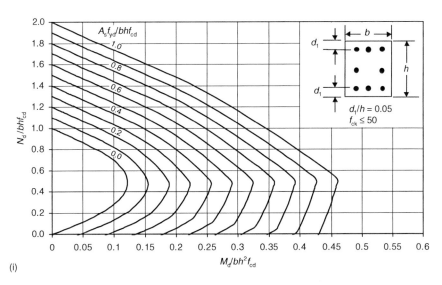

(i)

Fig. 5.8. (Contd)

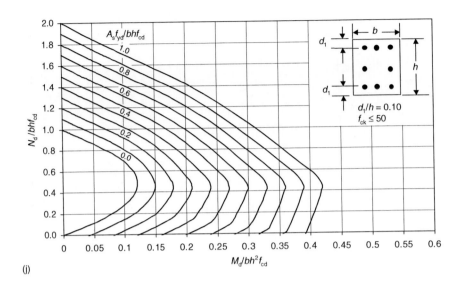

(j)

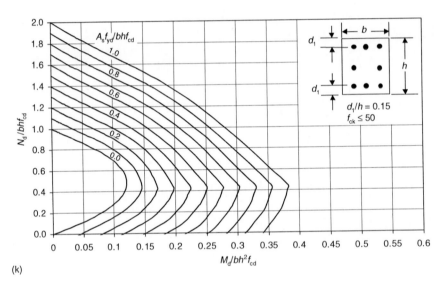

(k)

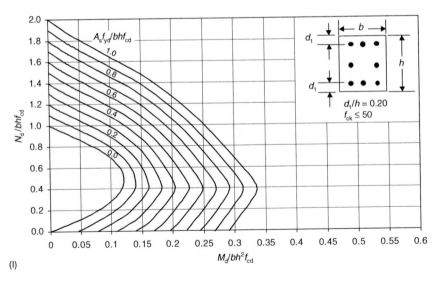

(l)

Fig. 5.8. (Contd)

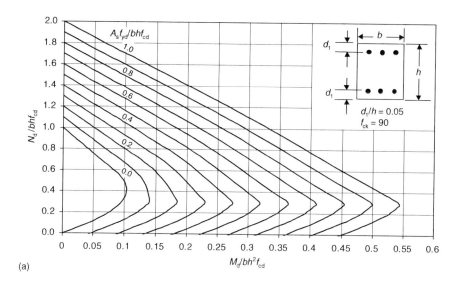

(a)

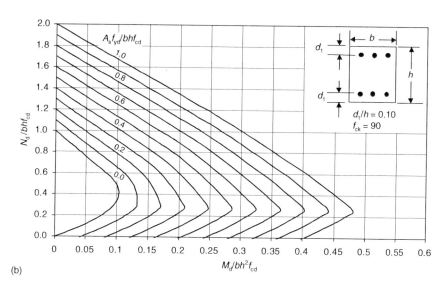

(b)

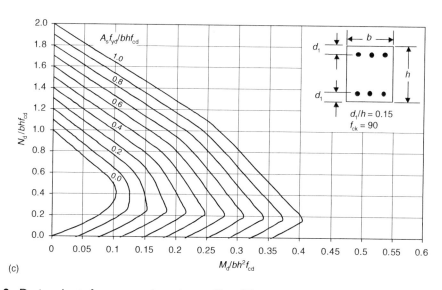

(c)

Fig. 5.9. Design charts for rectangular columns ($f_{ck} = 90$)

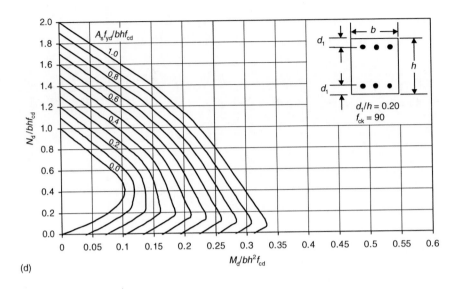

(d)

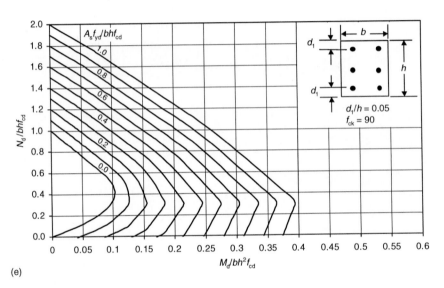

(e)

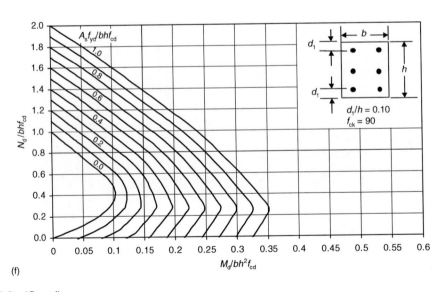

(f)

Fig. 5.9. (Contd)

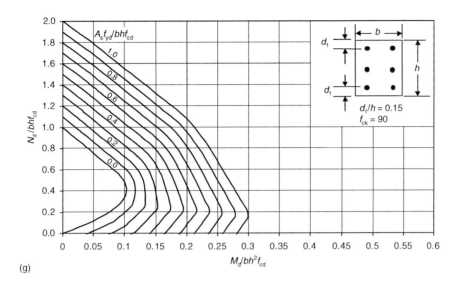

(g)

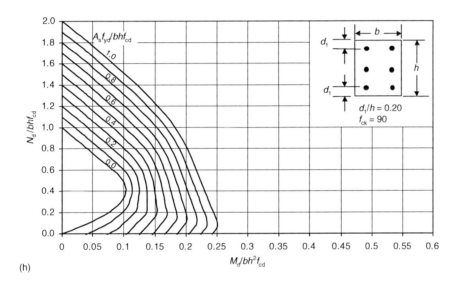

(h)

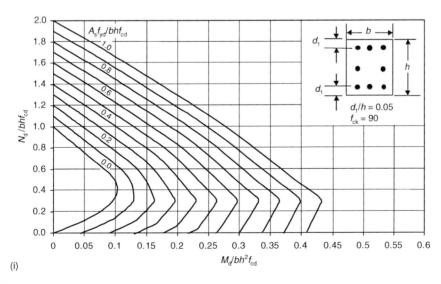

(i)

Fig. 5.9. (Contd)

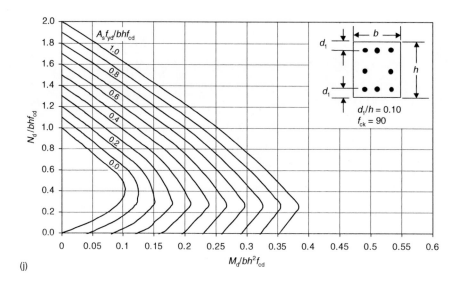

(j)

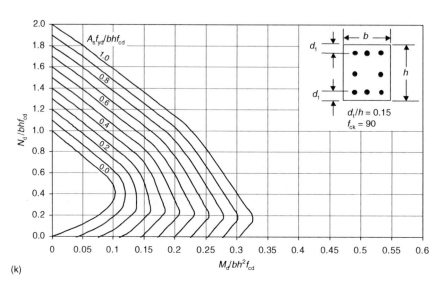

(k)

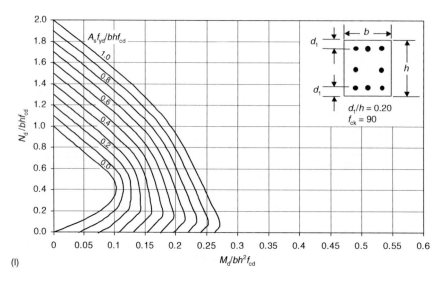

(l)

Fig. 5.9. (Contd)

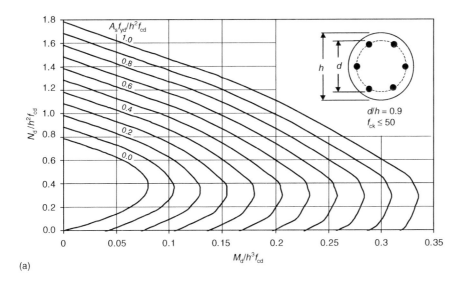

(a)

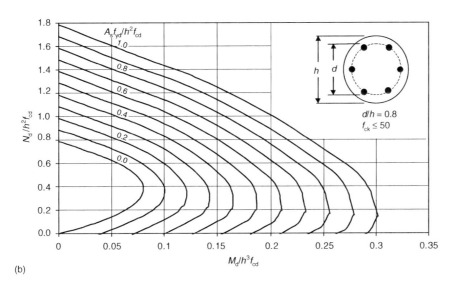

(b)

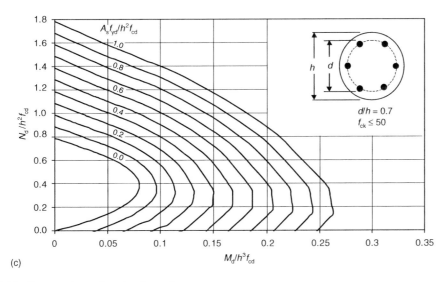

(c)

Fig. 5.10. Design charts for circular columns

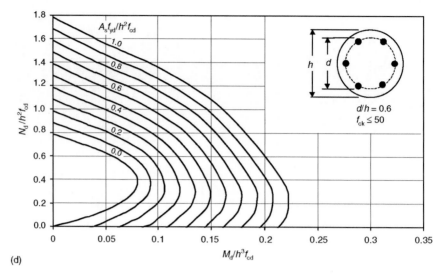

(d)

Fig. 5.10. (Contd)

Example 5.3: rectangular column section

A rectangular column section 500 mm deep and 300 mm wide is to be designed to carry a design ultimate axial load of 1875 kN and a design ultimate moment of 280 kN m. The characteristic strength of the reinforcement is 500 N/mm^2 and that of the concrete 30 N/mm^2.

The non-dimensional moment and axial load parameters are calculated as follows:

$f_{cd} = \alpha_{cc} f_{ck}/\gamma_m = 0.85 \times 30/1.5 = 17$

$f_{yd} = 500/1.15 = 435$

$N_d/bhf_{cd} = 1875 \times 1000/(300 \times 500 \times 17) = 0.735$

$M_d/bh^2 f_{cd} = 280 \times 10^6/(300 \times 500^2 \times 17) = 0.22$

Assuming that the centroid of the steel in each face is 50 mm from the face and that the reinforcement is concentrated at the corners of the section, Fig. 5.8b can be used to establish that the mechanical reinforcement ratio is 0.4. This gives a total steel area of

$0.40 \times 17 \times 500 \times 300/435 = 2345$ mm^2

5.2.6. Design for biaxial bending

EN 1992-1-1 does not directly give a method for designing biaxially bent columns other than working from first principles. This is not easy to do without design charts or a suitable computer program. In principle, the development of a suitable computer program is relatively simple, and the necessary assumptions are illustrated in Fig. 5.11. To develop the most rigorous program, it is convenient to divide the compression zone into strips parallel with the neutral axis and then calculate the stress in each strip using the parabolic–rectangular diagram. The force and the moment about the x and y axes of each strip at the ultimate limit state can then be calculated and summed to find the moment and axial force developed at the ultimate limit state by the concrete in compression. Such a program can be used to generate design charts for biaxial bending, if desired.

In this section, a number of simplified methods will be considered which will allow the design of rectangular sections to be carried out relatively simply.

Three methods of dealing with biaxially bent rectangular sections will be presented here. The methods are given in order of increasing simplicity and increasing approximation.

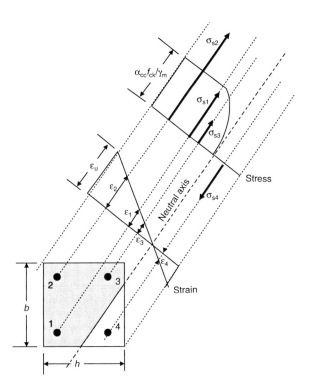

Fig. 5.11. Biaxial bending of a rectangular section

Table 5.4. Values of the exponent a in equation (D5.15)

N_{Ed}/N_{Rd}	a
≤ 0.1	1.00
0.7	1.5
1.0	2.0

The first is an approach given in *clause 5.8.9(4)* of EN 1992-1-1 for the design of biaxially bent sections in slender columns. This is based on the observation that the form of the M_x–M_y interaction diagram can conveniently be represented by a super-ellipse, and is believed to have been first developed by Bresler. A super-ellipse has an equation of the form   *Clause 5.8.9(4)*

$$x^a + y^a = k \qquad (D5.14)$$

If $a = 2$, this equation becomes a circle, while if $a = 1$ it describes a straight line. At loads approaching the squash load, the M_x–M_y interaction diagram approaches a circle, while in the region of the balance point it is close to a straight line. *Clause 5.8.9(4)* adopts the   *Clause 5.8.9(4)* equation below as a means of describing the complete interaction surface:

$$(M_{Edx}/M_{Rdx})^a + (M_{Edy}/M_{Rdy})^a = 1 \qquad (D5.15)$$

A convenient parameter for defining the proximity to the squash load is the ratio N/N_{uz}, and EN 1992-1-1 assumes the relationship between this parameter and the exponent a given in Table 5.4. Intermediate values may be interpolated.

In Table 5.4, N_{Rd} is the squash load of the column, and may be calculated from

$$N_{Rd} = A_c f_{cd} + A_s f_{yd}$$

The difficulty with the approach from the practical point of view is that it cannot be used as a direct design method since N_{Rd} can only be established once the reinforcement area has been found. It therefore has to be used iteratively. An initial estimate is made of N/N_{uz}, the section is designed, a corrected value of N/N_{uz} can then be estimated, and the process repeated until a correct solution is obtained.

It is not clear why the method described above is included only in the code for the design of slender columns, as it was originally devised for general use, and is considered to be adequately accurate for column sections where the reinforcement is concentrated near the corners. If the reinforcement is more uniformly distributed around the perimeter of the section, then the method is conservative.

A much simpler, but considerably more approximate, method has been adopted in BS 8110. This is a slightly modified version of a method given in *CEB Bulletin d'Information 141*.[8] Design is carried out for an increased uniaxial moment, which takes account of the biaxial effects. The required uniaxial moment is obtained from whichever is appropriate of the two relationships set out below:

$$\text{if } M_x/h' > M_y/b' \text{ then } M_x' = M_x + \beta h' M_y/b' \qquad (D5.16)$$

$$\text{if } M_x/h' < M_y/b' \text{ then } M_y' = M_y + \beta b' M_x/h' \qquad (D5.17)$$

In the above relationships, M_x and M_y are the design moments about the x and y axes, respectively, while M_x' and M_y' are the effective uniaxial moments for which the section is actually designed. b' and h' are effective depths as indicated in Fig. 5.12. The factor β is defined in BS 8110 as a function of N/bhf_{cu}. In terms of f_{ck}, it can be obtained from the relationship

$$\beta = 1 - N/bhf_{ck} \qquad (0.3 < \beta < 1.0) \qquad (D5.18)$$

This approach has the great advantage of being very simple. It is, however, an approximate approach, and it effectively defines the interaction diagram illustrated in Fig. 5.13. A possible way to use this simplified method is to use it to obtain a first approximation to the required area of reinforcement and then check the result using equation (D5.15).

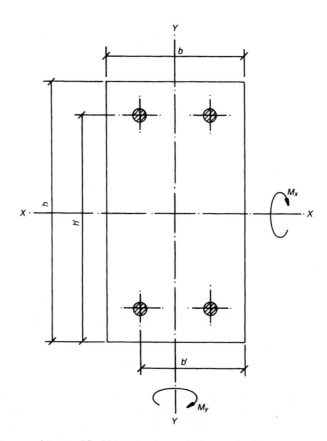

Fig. 5.12. Notation used in simplified biaxial column design method

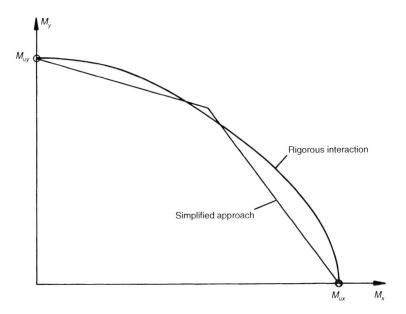

Fig. 5.13. Comparison of simplified and rigorous biaxial methods

Example 5.4: biaxially bent column section
A rectangular column section of dimensions 500 mm × 400 mm is to be designed to sustain a design axial load of 2000 kN combined with design moments of 100 kN m about the minor axis and 234 kN m about the major axis. The centres of the reinforcing bars may be assumed to be 50 mm from the shorter faces of the section and 60 mm from the longer faces. This gives $d'/h = 0.125$ and $d'/b = 0.2$. The concrete grade is C50/60, and the reinforcement strength is 460 N/mm². The reinforcement may be assumed to be concentrated close to the corners of the section.

The required area of reinforcement will be calculated using both the methods presented above, starting with the more exact method. However, analysis has been carried out using a rigorous method which has given a total area of reinforcement of 3216 mm². This value can be compared with the results obtained by the simplified methods to give some idea of their accuracy.

The simplest method (that from BS 8110) will be tried first.

The first step is to assess about which axis the effective uniaxial moment should be applied. $h' = 400 - 50 = 350$ and $b' = 300 - 60 = 240$. Using these values, $M_x/h' = 234/350 = 0.67$ and $M_y/b' = 100/240 = 0.417$. Since $M_x/h' > M_y/b'$, design is carried out for an effective uniaxial moment about the major (x) axis.

The coefficient β is given as

$1 - N_{Ed}/bhf_{ck} = 1 - 2\,000\,000/(300 \times 400 \times 50) = 0.6666$

The effective uniaxial moment is thus

$M'_x = M_x + \beta h'/b' M_y = 234 + 0.6666 \times 350 \times 100/240 = 331.2$ kN m

The design charts can now be used to obtain a steel area. It will be seen that it will be necessary to interpolate between the charts for bars in the corners with $d'/h = 0.1$ and $d'/h = 0.2$.

The parameterized axial force and moment are given by

$N_{Ed}/bhf_{cd} = 2\,000\,000/(400 \times 300 \times 50 \times 0.85/1.5) = 0.588$

$M'_x = 331.2 \times 10^6/(300 \times 400^2 \times 50 \times 0.85/1.5) = 0.244$

Figure 5.8b for $d'/h = 0.1$ gives $A_s f_{yd}/bhf_{cd} = 0.38$, while Fig. 5.8c for $d'/h = 0.15$ gives 0.44. This gives a value of 0.41 for $d'/h = 0.125$, and hence the area of reinforcement is given by

$$A_s = 0.41 \times 400 \times 300 \times 50 \times 0.85 \times 1.15/1.5/500 = 3206 \text{ mm}^2$$

(Note that $460/1.15 = 400$.) This can be seen to be almost indistinguishable from the rigorous answer of 3216 mm².

As mentioned above, the BS 8110 method is approximate, and the close result here may be the result of a fortunate chance rather than a general rule. It was suggested earlier that, having produced an approximate reinforcement area using the BS 8110 approach, it could then be refined by using the simplified method given in EN 1992-1-1. This will now be done.

Having obtained an approximate area of reinforcement, it is now necessary to calculate the moment of resistance of the section under uniaxial bending about the x and y axes. From the simplified calculation, we already know that the moment of resistance about the x axis of a column with 3206 mm² of reinforcement is 331.2 kN m. To find the moment of resistance for uniaxial bending about the minor axis, we need to use the chart for $d'/h = 0.2$. For $N_{Ed}/bhf_{cd} = 0.588$ and $A_s f_{yd}/bhf_{cd} = 0.41$, the chart gives $M/bh^2 f_{cd} = 0.215$. This gives a moment of

$$0.215 \times 400 \times 300^2 \times 50 \times 0.85/1.5/10^6 = 219 \text{ kN m}$$

The factor a must now be found.

$$N_{Rd} = A_c f_{cd} + A_s f_{yd} = (400 \times 500 \times 0.85 \times 50/1.5 + 3206 \times 500/1.15)/1000 = 4794 \text{ kN}$$

N_{Ed}/N_{Rd} is thus $2000/4794 = 0.417$. Interpolation of the table in *clause 5.8* of EN 1992-1-1 gives $a = 1.264$.

For the section to be satisfactory,

$$(M_{Edx}/M_{Rdx})^a + (M_{Edy}/M_{Rdy})^a \leq 1$$

Substituting the results:

$$(M_{Edx}/M_{Rdx})^a + (M_{Edy}/M_{Rdy})^a = (234/331.2)^{1.264} + (100/219)^{1.264} = 1.02$$

This is very close to 1.0, suggesting that the reinforcement area is close enough to the correct answer.

Thus, for this particular example, both the simplified methods give results which are indistinguishable from the rigorously calculated area of reinforcement.

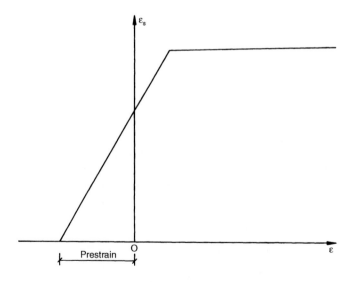

Fig. 5.14. Effective steel stress-strain curve for design of prestressed sections

5.2.7. Design of prestressed sections
The design of prestressed sections introduces no major new problems except that allowance must be made for the prestrain in the prestressing steel. This is effectively done by fixing a false origin for the stress–strain curve for the prestressing steel at a point corresponding to zero stress and a strain equal to the prestrain. This is illustrated in Fig. 5.14.

CHAPTER 6

Shear, punching shear and torsion

6.1. Shear – general

Shear is dealt with in *clause 6.2* of EN 1992-1-1, torsion in *clause 6.3* and punching shear in *clause 6.4*.

Clause 6.2
Clause 6.3
Clause 6.4

Probably more research has been carried out into shear behaviour than into any other mode of failure of reinforced or prestressed concrete. Despite this, there remain considerable areas of uncertainty and disagreement. Furthermore, unlike flexural behaviour, there is no generally accepted overall model describing shear behaviour. Despite these uncertainties, design for shear can be carried out with confidence for all normal members because the design methods given in codes have had the advantage of being tested against, and being adjusted to fit, a very large body of experimental data. The first part of this chapter will give an outline of shear behaviour as it is currently understood with an indication of how the provisions in the code reflect this.

6.2. Background to the code provisions

Four aspects of the strength of sections subjected to shear loading will be considered here. These are:

- strength of members without shear reinforcement
- strength of members with shear reinforcement
- maximum shear that can be carried by a member
- behaviour close to supports.

Each of these will be considered in turn.

6.2.1. Members without shear reinforcement

This area is probably of limited importance for beams where some shear reinforcement will always be provided but is of major importance for slabs where it is often very inconvenient to provide shear reinforcement. It is of particular importance for design for punching (see Section 6.5). This has been the most researched area of shear performance, but, as yet, no generally accepted theory describing the ultimate behaviour of a member without shear reinforcement has been developed. The formulae given in codes should therefore be considered to be basically empirical. Because of the amount of testing that has been carried out, the effect of the major variables can be clearly established, and the resulting formulae can be considered as highly reliable for normal types of element.

The principal variables governing the shear strength of members without shear reinforcement are the concrete strength, the reinforcement ratio and the depth of the member.

It is fairly self-evident that concrete strength will have an influence on shear strength, though the nature of the failure suggests that it is likely to be the tensile strength rather than the compressive strength which is important. In fact, experimental evidence suggests that the influence of concrete strength is rather less than proportional to the tensile strength. EN 1992-1-1 gives the tensile strength as varying in proportion to the compressive strength to the power of 2/3. The code indicates the shear strength of sections unreinforced in shear to vary in proportion to the compressive strength to the power of 1/3. In this it agrees with UK practice as incorporated in BS 8110.

Shear strength increases with increasing reinforcement ratio, but the rate of increase reduces as reinforcement ratio increases. This behaviour may be modelled in various ways. EN 1992-1-1, like BS 8110, employs a cube root relationship between shear strength and steel percentage up to a maximum value of ρ of 0.02.

The absolute section depth is found to have a significant influence on shear strength over and above that expected from normal geometrical scaling (i.e. there is a size effect). This effect is sufficiently large that it is worth taking into account in the design of shallow members such as slabs. Most recent codes of practice therefore have a term in their equations to allow for this and which gives a higher shear strength for shallow members.

Taking these various factors into account, EN 1992-1-1 gives the following equation for the strength of sections without shear reinforcement:

$$V_{\text{Rd, c}} = b_{\text{w}}d[(0.18/\gamma_{\text{c}})k(100\rho_1 f_{\text{ck}})^{1/3} + 0.15\sigma_{\text{cp}}]$$ (D6.1)

$V_{\text{Rd, c}}$ should not be taken as less than

$$b_{\text{w}}d(0.035k^{3/2}f_{\text{ck}}^{1/2} + 0.15\sigma_{\text{cp}})$$ (D6.2)

where

$$k = 1 + \sqrt{(200/d)} \le 2.0$$

d is in millimetres,

$$\rho_1 = A_{\text{sl}}/b_{\text{w}}d \le 0.02$$

and σ_{cp} is the average longitudinal stress (note compression is positive). The tensile axial force is simply taken into account by the introduction of a negative value for σ_{cp}.

In equation (D6.1), the term $0.18/\gamma_{\text{c}}$ and the constant 0.15 are nationally determined parameters. The UK has, however, accepted the recommended values, and these are as given above. For convenience, they have been included explicitly in the equation.

The minimum requirement for $V_{\text{Rd, c}}$ takes account of the fact that a member without any reinforcement still has some shear strength.

6.2.2. Strength of sections with shear reinforcement

There is a generally accepted model for the prediction of the effects of shear reinforcement. This is the 'truss model', illustrated in Fig. 6.1 for the commonest case where vertical links are used. In this model, the top and bottom compression and tension members are, respectively, the concrete in the compression zone and the tension steel. The members connecting the top and bottom members are represented by steel tension members and virtual concrete 'struts'. The truss in Fig. 6.1 can be analysed to give the forces in the various members as indicated below:

$$F_1 = N/2 + V(a - v/z - 0.5 \cot \theta)$$ (D6.3)

$$F_3 = N/2 - V(a - v/z + 0.5 \cot \theta)$$ (D6.4)

$$F_2 = V/\sin \theta$$ (D6.5)

$$F_4 = V$$ (D6.6)

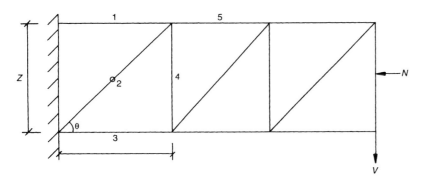

Fig. 6.1. Idealized truss

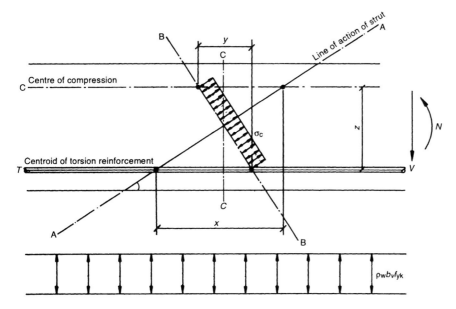

Fig. 6.2. Idealization of a beam with shear reinforcement as a 'smeared' truss

Two factors can be observed from this analysis. Firstly, the forces in the vertical tension member 4 and the compression strut are both independent of the axial force, N. Secondly, the forces in the compression and tension chords both differ from that calculated from the moment alone by $0.5V \cot \theta$. (Note that $Va_v = M$, and the forces in the upper and lower chords are given by bending theory as $\pm M/z$.)

In developing the truss under discussion into a design approach for a reinforced or prestressed concrete beam, it is necessary to consider it as a 'smeared' truss, as illustrated in Fig. 6.2. In this idealized truss, the vertical stirrups are represented by a uniform vertical tensile stress of $\rho_w f_{yk}$ or a vertical force of $\rho_w f_{yk} b_w$ per unit length of the beam. This assumes that, at failure, the stirrups yield, The virtual compressive strut is replaced by a uniform uniaxial compressive stress of σ_c acting parallel to the line of action of the strut over the concrete between the centroid of the tension reinforcement and the centre of compression. A complete analysis of this smeared truss may be carried out by considering the equilibrium of two sections:

*   Vertical equlibrium across a section parallel to the virtual compression strut. In this section, all the shear is supported by the shear reinforcement (section A–A in Fig. 6.2). This gives

$$V = \rho_w f_{yk} b_w z \cot \theta \qquad (D6.7)$$

- Vertical equilibrium across a section perpendicular to the strut (section B–B in Fig. 6.2). Across this section, some of the shear is carried by the strut and some by the stirrups. The total strut force is

$$F_c = \sigma_c b_w z/\cos \theta \tag{D6.8}$$

The horizontal projection of the section is given by $z \tan \theta$, and hence equilibrium is given by

$$V = \rho_w f_{yk} b_w z \tan \theta + F_c \cos \theta$$

Since, from equation (D6.7),

$$\rho_w f_{yk} b_w z = V \tan \theta$$

we can rewrite this as

$$V = V \tan^2 \theta + \sigma_c b_w z \tan \theta$$

This can be rearranged to give

$$V = \sigma_c b_w z/(\cot \theta + \tan \theta) \tag{D6.9}$$

These equations can be converted to design values by simply replacing the characteristic material properties by the appropriate design values. σ_c is replaced by $\alpha_{cw} \nu_1 f_{cd}$. The parameter ν_1 is an efficiency factor which allows for the actual distribution of the stress within the strut and the effects of cracking. α_{cw} is a coefficient taking account of any applied compression force. Both these factors are Nationally Defined Parameters, and the recommended value for ν_1 is ν given by

$$\nu = 0.6(1 - f_{ck}/250) \tag{D6.10}$$

A higher value is permitted where the design stress in the shear reinforcement is less than $0.8f_{yk}$.

α_{cw} takes a value of 1 for non-prestressed structures or structures subjected to tension. For cases where an axial compression is applied, *equations (6.11.aN)*, *(6.11.bN)* and *(6.11.cN)* in EN 1992-1-1 provide values.

Equation (D6.9) then gives the maximum shear that can be carried by a section before failure by crushing of the notional compression struts.

Equations for inclined shear reinforcement can be derived in a similar way by replacing the vertical tension by a uniform tension inclined at an angle to the horizontal.

The design forms of equations (D6.7) and (D6.9) appear in EN 1992-1-1 as *equations (6.8)* and *(6.9)*.

There are two areas where methods in various codes differ. The first is in the choice of a value for the truss angle, θ. The second is whether the truss should carry all the shear force or whether part of the shear can be considered to be carried by the concrete.

Internationally, by far the most common approach is to assume that $\cot \theta$ is 1 (i.e. the truss is assumed to form at 45° to the axis of the beam). When this assumption is made, it is found that the shear capacity of a beam with shear reinforcement is underestimated, and it is found that a generally consistent prediction of strength is achieved if the shear strength is assumed to be equal to the shear capacity of the shear reinforcement calculated according to equation (D6.9) plus the capacity of the concrete given by equation (D6.1) (or its equivalent in whatever code is being considered). The shear response is thus assumed to be plastic. Both BS 8110 and the ACI code follow this approach.

EN 1992-1-1 adopts a second approach. In this approach, all the shear is assumed to be carried by the shear reinforcement, but the truss angle, θ, can take any value between $\cot^{-1} 0.4$ and $\cot^{-1} 2.5$. This variable strut angle approach is considered to be the more rigorous of the two methods, and is also slightly more economical in some cases. However, as written in EN 1992-1-1, it is open to a misunderstanding. The code implies that the designer may select any strut angle between the specified limits. This concept of free choice does not,

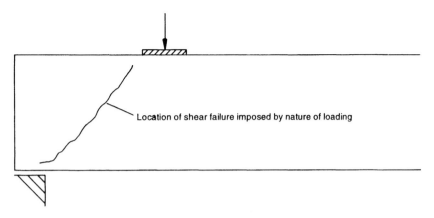

Fig. 6.3. Shear failure close to a support

however, reflect the behaviour of a beam. Beams will fail in a manner corresponding to a strut angle of roughly $\cot^{-1} 2.5$ unless constrained by the detailing or the geometry of the system to fail at some steeper angle. A steeper-angled failure could be induced either by the way in which the tension steel was curtailed or where the load is so close to the support that only a steeper failure can occur or where the shear strength is limited by the crushing strength of the strut. This last consideration will be considered further after the crushing strength of the strut has been discussed. Figure 6.3 illustrates this aspect of behaviour.

6.2.3. Maximum shear strength of a section

The previous section has been concerned with the behaviour of the tension members in the truss. The capacity of these can be increased to whatever level is required simply by increasing the amount of reinforcement provided. The capacity of the compression members, the virtual diagonal struts, cannot, however, be so easily varied. It is the strength of these struts that provides an absolute upper limit to the shear that can be supported by the beam. The shear supported by these members has been derived above as

$$V = \sigma_c b_w z / (\cot \theta + \tan \theta)$$

At the ultimate load, the strut will crush, and the average stress in the strut, σ_c, can be expected to be proportional to the compressive strength of the concrete. This is achieved by defining the limiting value of σ_c as νf_{cd}, where ν is an empirically obtained efficiency factor which can be considered to take account of the actual distribution of stress across the section at ultimate. Thus:

$$V_{Rd,\,max} = b_w z \alpha_{cw} \nu_1 f_{cd} / (\cot \theta + \tan \theta) \tag{D6.11}$$

where $V_{Rd,\,max}$ is the maximum shear capacity of the section. EN 1992-1-1 states that z may generally be assumed to be $0.9d$.

These equations are appropriate for vertical shear reinforcement.

It will be noted from equation (D6.7) that the smaller the angle θ, the greater is the shear capacity based on the shear reinforcement. However, the shear capacity based on the crushing strength of the strut, given by equation (D6.11), decreases with decreasing values of θ below 45°. Hence the maximum capacity corresponds to the situation where the capacity based on the shear reinforcement just equals the capacity based on the strength of the strut. This implies that the actual conditions at failure may be established by using equation (D6.11) to estimate the value of θ for which $V = V_{Rd,\,max}$, and then using this value of θ to obtain the required amount of shear reinforcement. It should be noted that $V_{Rd,\,max}$ reaches a maximum value when $\cot \theta = 1$, and hence values of θ greater than 45° will only occur if other factors constrain the failure to occur at such an angle. The result of combining the strength defined by the crushing strength of the strut and the limitations applied to the strut angle is illustrated in Fig. 6.4.

6.2.4. Shear capacity enhancement near supports

There is very extensive test work that shows that much higher shear strengths than are given by equation (D6.1) can be obtained in short members such as corbels or in beams where the load is applied close to the support. The reason for this is simply that, for any sections closer to the support than the critical section defined by the 'natural' strut angle of $\cot^{-1} 2.5$ (Fig. 6.5), a substantial proportion of the load will be carried through to the support directly by the strut and not via the normal actions of shear and bending. The closer the load is to the support, the greater the proportion of the load which will be transmitted to the support in this way. Experimental evidence suggests that a convenient method of treating this problem is to assume that the shear carried by this mechanism is a function of d/x, where x is the distance from the face of the support to the face of the load (see Fig. 6.5). Figure 6.6 shows experimental results for failures close to supports. EN 1992-1-1 deals with this problem by increasing the design shear resistance of the concrete section, $V_{Rd,c}$, according to the following modified equation:

$$V_{Rd,c} = b_w d[(0.18/\gamma)k(100\rho_1 f_{ck})^{1/3}(2d/x) - 0.15\sigma_{cp}] \leq 0.5 b_w d\nu f_{cd} \qquad (D6.12)$$

EN 1992-1-1 does not mention the shear reinforcement that would be required in circumstances where this was required at sections closer than $2d$ to the face of a support. Clearly, in such cases the truss angle θ is constrained so that $\cot \theta$ at the section considered is not greater than $x/0.9d$.

A further point that should be noted is that, clearly, this enhancement can only be applied where the load is applied to the top face of the beam and the support is at the bottom,

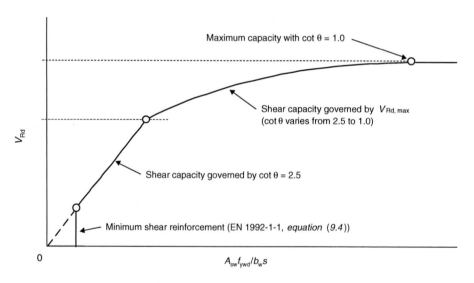

Fig. 6.4. Relationship between the design shear force and the amount of shear reinforcement

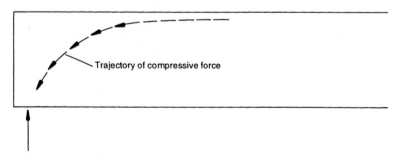

Fig. 6.5. Direct transfer of the load to the support

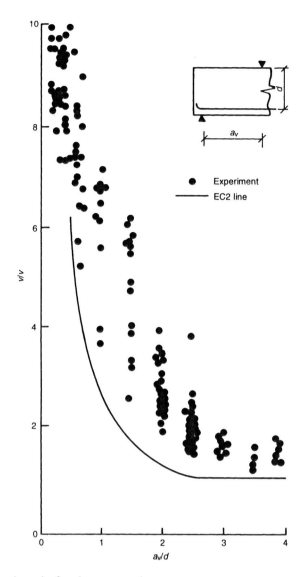

Fig. 6.6. Experimental results for shear strength near supports

and where the longitudinal reinforcement is properly anchored beyond the centreline of the support.

6.2.5. Summary
This section has tried to give a brief picture of the phenomenon of shear failure and the factors which influence it. This should help in understanding the provisions of EN 1992-1-1 and in applying them with confidence. It should be understood, however, that this picture is a simplified one: shear is still not fully understood, and remains a very complex phenomenon.

6.3. Summary of the provisions in *clause 6.2* of EN 1992-1-1
Tables 6.1–6.3 will enable the basic quantities used in the shear calculation to be read off for most normal situations which will be met with for reinforced concrete.

Table 6.1 can be used to obtain basic values of $V_{Rd, c}/b_w d$ for different values of longitudinal reinforcement ratio and concrete grade. These values have been calculated ignoring the factor k. To obtain the correct value for $V_{Rd, c}/b_w d$, the values will need to be multiplied by the

k factor given in Table 6.2. Table 6.3 then gives the minimum value of $V_{Rd, c}/b_w d$. For members supporting an axial force (i.e. a column or a prestressed beam), the value of $V_{Rd, c}$ obtained from Tables 6.1 and 6.3 are increased by adding a shear equal to $0.15N/(A_c b_w d)$, where N is the axial force and A_c is the cross-sectional area of the element.

Where shear reinforcement in the form of vertical stirrups is to be provided, Fig. 6.7 may be used to calculate the required amount. To use the chart, first estimate the value of $V_{Ed}/b_w d$. The design relationship is then defined by two curves: (1) a straight line defined by

Table 6.1. Values of $(0.18/\gamma_c)(100\rho_1 f_{ck})^{1/3}$

Longitudinal reinforcement (%)	Concrete compressive strength, f_{ck} (N/mm^2)										
	25	30	35	40	45	50	55	60	70	80	90
0.2	0.21	0.22	0.23	0.24	0.25	0.26	0.27	0.27	0.29	0.30	0.31
0.4	0.26	0.27	0.29	0.30	0.31	0.33	0.34	0.35	0.36	0.38	0.40
0.6	0.30	0.31	0.33	0.35	0.36	0.37	0.38	0.40	0.42	0.44	0.45
0.8	0.33	0.35	0.36	0.38	0.40	0.41	0.42	0.44	0.46	0.48	0.50
1.0	0.35	0.37	0.39	0.41	0.43	0.44	0.46	0.47	0.49	0.52	0.54
1.2	0.37	0.40	0.42	0.44	0.45	0.47	0.48	0.50	0.53	0.55	0.57
1.4	0.39	0.42	0.44	0.46	0.48	0.49	0.51	0.53	0.55	0.58	0.60
1.6	0.41	0.44	0.46	0.48	0.50	0.52	0.53	0.55	0.58	0.60	0.63
1.8	0.43	0.45	0.48	0.50	0.52	0.54	0.56	0.57	0.60	0.63	0.65
2.0	0.44	0.47	0.49	0.52	0.54	0.56	0.57	0.59	0.62	0.65	0.68

Table 6.2. Values of depth factor, k

Effective depth (mm)	k
100	2.00
200	2.00
300	1.82
400	1.71
500	1.63
600	1.58
700	1.53
800	1.50
900	1.47
1000	1.45

Table 6.3. Values of v_{min}

Effective depth (mm)	Concrete compressive strength, f_{ck} (N/mm^2)										
	25	30	35	40	45	50	55	60	70	80	90
100	0.67	0.73	0.79	0.85	0.90	0.95	0.99	1.04	1.12	1.20	1.27
200	0.35	0.38	0.41	0.44	0.47	0.49	0.52	0.54	0.59	0.63	0.66
300	0.27	0.30	0.32	0.34	0.36	0.38	0.40	0.42	0.45	0.48	0.51
400	0.24	0.26	0.28	0.30	0.32	0.33	0.35	0.37	0.40	0.42	0.45
500	0.22	0.24	0.26	0.28	0.29	0.31	0.33	0.34	0.37	0.39	0.42
600	0.21	0.23	0.25	0.26	0.28	0.30	0.31	0.32	0.35	0.37	0.40
700	0.20	0.22	0.24	0.26	0.27	0.29	0.30	0.31	0.34	0.36	0.38
800	0.20	0.22	0.23	0.25	0.26	0.28	0.29	0.30	0.33	0.35	0.37
900	0.19	0.21	0.23	0.24	0.26	0.27	0.29	0.30	0.32	0.35	0.37
1000	0.19	0.21	0.23	0.24	0.26	0.27	0.28	0.30	0.32	0.34	0.36

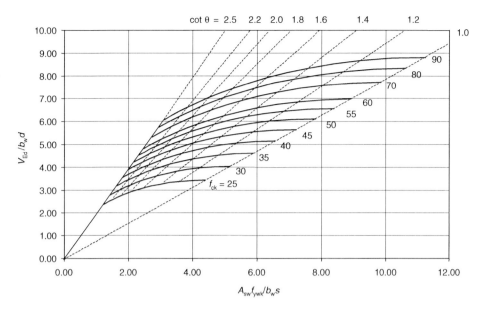

Fig. 6.7. Design chart for the selection of shear reinforcement

the maximum permissible value of cot θ (cot $\theta = 2.5$) and (2) a curve defined by the grade of concrete. The broken lines on the chart give the values of cot θ. These may be useful when detailing the longitudinal reinforcement, as the curtailment depends to some degree on the value of cot θ. The right-hand broken line, defining the end of the curves, corresponds to cot $\theta = 1$ and gives the maximum amount of vertical shear reinforcement that may be used. Minimum areas of shear reinforcement are not marked on Fig. 6.7, as this would overcomplicate the chart; however, values can be obtained from *equation (9.5)* in EN 1992-1-1:

$$(A_{sw}f_{ywk}/b_w s)_{min} = 0.08\sqrt{f_{ck}} \qquad \text{(D6.13)}$$

In selecting the critical section for shear design, it should be noted that *clause 6.2.2(3)* in EN 1992-1-1 indicates that the shear resistance need not be checked closer to the support than the point where a line drawn at 45° from the inner face of the support intersects the elastic centroidal axis of the member. This takes account of the increased shear resistance near the supports. For concentrated loads, this effect is accounted for by the shear force reduction factor given earlier.

Clause 6.2.2(3)

It should be noted again that all the equations derived and presented in this chapter deal only with the case of vertical shear reinforcement. No new principles are involved in extending the derivations to the case of inclined shear reinforcement, it simply complicates the equations and makes the derivation less clear. Furthermore, from the practical point of view, the use of inclined shear reinforcement is becoming increasingly rare due to the considerably increased expense of the reinforcement fixing and difficulties in maintaining tolerances. Should inclined shear reinforcement be needed, the necessary equations are all presented in EN 1992-1-1 itself.

The best way to illustrate the design procedure is by means of examples.

The shear resistance of a section without shear reinforcement is given by

$$V_{Rd,c} = [(0.18/\gamma_c)k(100\rho_1 f_{ck})^{1/3} + 0.15\sigma_{cp}]b_w d \geq (v_{min} + 0.15\sigma_{cp})b_w d$$

Example 6.1: T section

The simply supported T beam illustrated in Fig. 6.8 is to be designed for shear. The beam is designed to use concrete grade C30/37 and reinforcement with a characteristic strength of 500 N/mm². The design for flexure has resulted in a requirement for 2200 mm² of

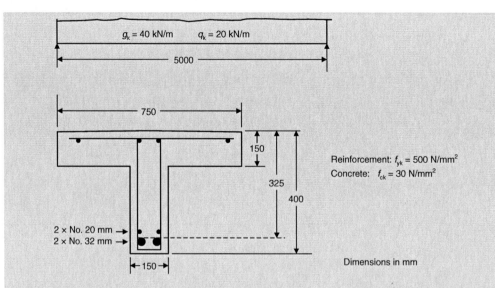

Fig. 6.8. Shear design example

longitudinal reinforcement. This is provided by $2 \times$ No. 32 mm and $2 \times$ No. 20 mm bars, giving an area of 2236 mm². The effective depth of the section is 325 mm, and the depth from the top of the flange to the elastic centroid of the section is 125 mm.

The shear force at the supports is given by

$$(1.35 \times 40 + 1.5 \times 20) \times 5/2 = 210 \text{ kN}$$

However, the design shear may be taken as that at a distance from the face of the support equal to the height from the bottom of the beam to the elastic centroid of the section from the face of the support. Assuming that the distance from the support centreline to the face of the support is 75 mm, the critical design shear is that at a distance of

$$75 + 400 - 125 = 350 \text{ mm}$$

from the support centreline. This shear can be calculated as

$$V_{\text{Ed}} = 210 \times (5000/2 - 350)/(5000/2) = 180.6 \text{ kN}$$

$V_{\text{Rd, c}}$ is now calculated. The reinforcement ratio is given by

$$A_{\text{sl}}/b_{\text{w}}d = 2236/150/325 = 4.59\%$$

However, the maximum value that may be used in assessing $V_{\text{Rd, c}}$ is 2%. From Table 6.1, the basic value of $V_{\text{Rd, c}}/b_{\text{w}}d$ is 0.47 N/mm², and from Table 6.2, the correction factor for depth is 1.78. Hence,

$$V_{\text{Rd, c}} = 0.44 \times 1.78 \times 150 \times 325/1000 = 40.8 \text{ kN}$$

This is much less than V_{Ed}, and hence designed shear reinforcement is needed:

$$V_{\text{Ed}}/b_{\text{w}}d = 180.6 \times 1000/150/325 = 3.69$$

Figure 6.7 gives the required value of $A_{\text{sw}}f_{\text{ywk}}/b_{\text{w}}s$ as 2.5. Hence,

$$A_{\text{sw}}/s = 2.5 \times 150/500 = 0.75$$

The maximum spacing of links according to *equation (9.6)* in EN 1992-1-1 is

$$0.75d = 0.75 \times 325 = 244 \text{ mm}$$

The most economical solution which will not violate the maximum stirrup spacing rules would thus seem to be to use 10 mm diameter stirrups at 200 mm centres.

The minimum permissible value of $A_{sw}f_{ywk}/b_w s$ is

$0.08 \times \sqrt{30} = 0.438$

giving $(A_{sw}/s)_{min}$ as

$0.438 \times 150/500 = 0.131$

Assuming that a bar diameter less than 8 mm will not be used, and recalling that the maximum bar spacing is 244 mm suggests that minimum links have to be 8 mm links at 225 mm centres, giving $A_{sw}/s = 0.447$. The shear capacity of this choice of minimum stirrups can be calculated from equation (D6.8) (noting that, in this case, $\cot\theta$ will be 2.5) as

$V_{Rd,s} = 0.447 \times 0.9 \times 325 \times 500 \times 2.5/1.15/1000 \text{ kN} = 142 \text{ kN}$

This indicates that minimum stirrups will be adequate over the central 3.4 m of the span, with the heavier designed shear reinforcement near the ends.

6.4. Torsion

6.4.1. Introduction

In normal building structures, torsion generally arises as a secondary effect, and specific calculations are not necessary. Torsional cracking is generally adequately controlled by reinforcement provided to resist shear. Even when torsion occurs, it rarely controls the basic sizing of members, and torsion design is often a check calculation after the members have been designed for flexure. Also in some of these cases, the loading that causes the maximum torsional moments may not be the same that induces the maximum flexural effects. In some cases, reinforcement provided for flexure and other forces may prove adequate to resist torsion.

It is important to recognize two basic types of torsion:

- Equilibrium torsion, which is essential for the basic stability of the element or structure, e.g. a canopy cantilevering off an edge beam (Fig. 6.9a). Equilibrium torsion is clearly a fundamental design effect, which cannot be ignored. Structures in which torsion design is carried out to include beams curved on plan, helical staircases and box girders.
- Compatibility torsion, which arises in monolithic construction when compatibility of deformation of the connected parts have to be maintained, e.g. edge beams in a normal framed building with slabs and beams framed in on one side only. In these cases the torsional restraint can be released without causing a collapse of the structure; however, at serviceability, cracking may occur in the absence of sufficient reinforcement (Fig. 6.9b).

6.4.2. Evaluation of torsional moments

When torsion has to be taken into account in design of framed structures, the torsional moments may be obtained on the basis of an elastic analysis using torsional rigidity (GC) calculated as follows:

$G = 0.42E_{cm}$

$C = 0.5 \times$ St Venant torsional stiffness for a plain concrete section

(Note: 0.5 has been justified by test results.)

For a rectangular section, the St Venant torsional stiffness is $k_1 b^3 d$, where k_1 is obtained from Table 6.4, b is the overall breadth ($< d$) and d is the overall depth.

Non-rectangular sections should be divided into a series of rectangles, and their torsional stiffness should be summed up. The division should be carried out such as to maximize the stiffness.

6.4.3. Verification

In EN 1992-1-1, the torsional resistance is calculated on the basis of a thin-walled closed cross-section. Solid sections are replaced by equivalent thin-walled sections. The wall thickness is taken as A/u, where A is the total area of cross-section within the outer circumference including inner hollow areas and u is the outer circumference of the cross-section. The effective thickness is to be not less than twice the distance between the outer edge and the centreline of the longitudinal reinforcement. In hollow sections; the effective

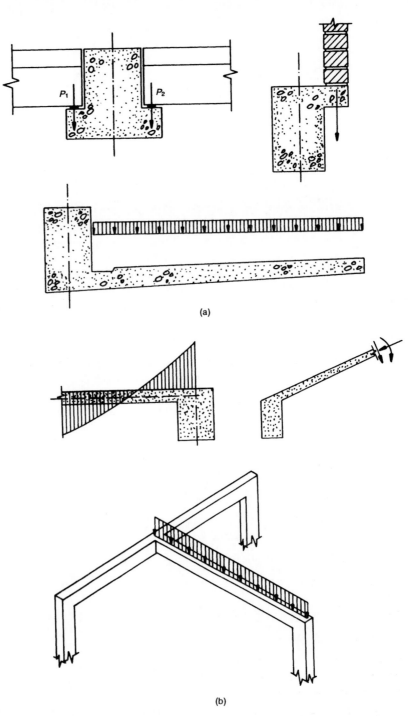

(a)

(b)

Fig. 6.9. Examples of (a) primary or equilibrium torsion and (b) torsion in statically indeterminate structures

Table 6.4. Values of k_1 for a rectangular section

d/b	k_1
1	0.14
1.5	0.20
2	0.23
3	0.26
5	0.26
> 5	0.33

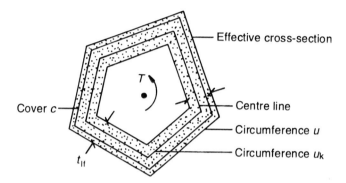

Fig. 6.10. Notation for torsion

thickness is not to be taken as greater than the actual wall thickness. Figure 6.10 defines the relevant parameters.

The procedure for verification may be summarized as follows:

(1) Choose an angle θ for the struts such that the following relationship is satisfied:

$$T_{Ed} = T_{Rd,\,max} = 2\nu f_{cd} A_k t_{ef} \sin\theta \cos\theta$$

in which T_{Ed} is the applied torsion and $\nu = 1 - (f_{ck}/250)$; A_k is the area enclosed by the centre lines of the connecting walls including the inner hollow areas, $\cot\theta$ should lie in the range 1.0 and 2.5. (Note that the above expression has to be multiplied by a factor α_{cw} if axial stresses exist. See EN 1992-1-1 for details.)

(2) The transverse reinforcement required should be calculated using the following expression:

$$A_{sw}/s = T_{Ed}/2A_k f_{yw,\,d} \cot\theta$$

where $f_{yw,\,d}$ is the design strength of the transverse reinforcement.

(3) The longitudinal reinforcement required to resist torsion is given by

$$A_{sl} = T_{Ed} u_k \cot\theta/2A_k f_{yd}$$

in which f_{yd} is the design strength of the longitudinal reinforcement and u_k is the perimeter of area A_k.

(4) In a member subject to torsion and shear, the following inequality should be satisfied:

$$T_{Ed}/T_{Rd,\,max} + V_{Ed}/V_{Rd,\,max} \leq 1.0$$

in which V_{Ed} is the applied shear force and $V_{Rd,\,max}$ is the maximum shear resistance of the cross-section in accordance with equation (D6.11).

(5) Only minimum reinforcement is required when

$$T_{Ed}/T_{Rd,\,c} + V_{Ed}/V_{Rd,\,c} \leq 1.0$$

in which $T_{Rd,\,c} = 2A_k t_{ef} f_{ct,\,d}$ and $V_{Rd,\,c}$ is as defined in equation (D6.1).

6.5. Punching shear

6.5.1. General

Punching shear is a local shear failure around a concentrated load on a slab. The most common situations where punching shear has to be considered is the region immediately surrounding a column in a flat slab or where a column is supported on a pad footing or foundation raft. Design for punching shear is covered in *clause 6.4* of EN 1992-1-1.

Clause 6.4

Punching shear failures may be considered to be shear failures rotated around the loaded area so as to give a failure surface which has the form of a truncated cone. This is illustrated in Fig. 6.11. The 'critical section' for shear failure in a beam is transformed into a 'basic control perimeter' when punching shear is considered. This conversion of the problem from a basically two-dimensional one to a three-dimensional problem does not change the basic phenomenon as described in Section 6.2, though there are a number of practical points which need further consideration. The first point which will be considered is the definition of the control perimeter.

6.5.2. Basic control perimeter

The starting point for design for punching shear is the definition of the critical perimeter. This has greater importance than the selection of the critical section in a beam because, as perimeters closer to the loaded area are considered, the length of the perimeter rapidly gets shorter, and hence the shear force per unit length of the perimeter rapidly increases. Observation of failures shows that the outer perimeter of a punching failure takes the general form sketched in Fig. 6.12a. For this reason, EN 1992-1-1 proposes the idealized form shown in Fig. 6.10b. Some national codes (e.g. the UK code, BS 8110) employ a rectangular perimeter. This is less realistic but slightly easier to use. Another advantage of the rectangular perimeter is that, when reinforcement is needed, this must be provided within the perimeter. It is much easier to do this with a rectangular perimeter since reinforcement is commonly provided on a rectangular grid. In fact, as will be seen below,

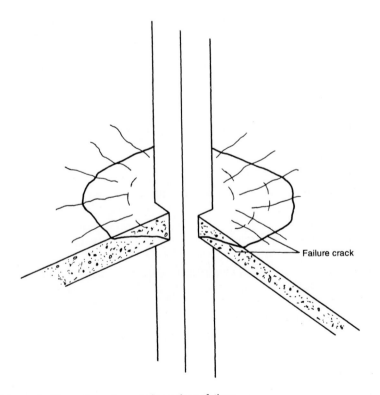

Fig. 6.11. Schematic illustration of a punching shear failure

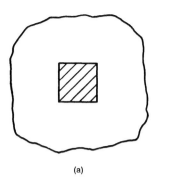 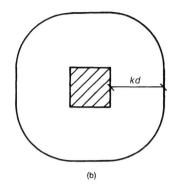

(a)   (b)

Fig. 6.12. Rationalisation of failure perimeter

since the design equations for punching shear are basically empirical, the precise choice of the shape of the perimeter in a particular code is not too important.

Having chosen a basic form for the perimeter, it is next necessary to consider the distance from the loaded area at which it should be located. Since the treatment of shear is empirical, there is no absolutely defined answer to this question. The same strength could be obtained by using a high value for the shear strength combined with a short perimeter close to a column or by using a lower shear strength combined with a longer perimeter further from the column. Different codes have adopted different solutions to this question; the UK code takes a perimeter 1.5d from the face of the column or loaded area while the US ACI code uses a perimeter 0.5d from the face of the column but with a much higher stress. In drafting EN 1992-1-1, it was decided that the shear strength should be given by the same formula for punching shear as is used for shear in beams. Having decided on this and on the shape of the perimeter, the distance of the perimeter from the loaded area can be established from test data. This led to a value of 2d. The interpretation of this to define the actual perimeter is dealt with very thoroughly in *clauses 6.4.2(1)–6.4.2(6)* of EN 1992-1-1, and needs no further discussion here.

Clauses 6.4.2(1)– 6.4.2(6)

A question which does require some further clarification is the definition of the control perimeter in cases where there is a 'drop' or 'column head' around the column. This is covered in EN 1992-1-1 in *clauses 6.4.2(8)* and *6.4.2(9)*] and *Figs 6.17* and *6.18*, but is not easy to follow. Although the code does not explicitly say so, it appears that, where there is a drop, a circular control perimeter is assumed with a radius r_{cont}. The provisions in *clauses 6.4.2(8)* and *6.4.2(9)* are concerned with defining this radius. The flow chart in Fig. 6.13 is intended to clarify the interpretation of these provisions.

Clause 6.4.2(8)
Clause 6.4.2(9)

6.5.3. Design shear force

It is usual to assume in design that the distribution of shear force around the critical perimeter is uniform. In fact, this is untrue, particularly in the case of a slab–column connection where there is a moment transfer between the slab and the column. In such cases, a rigorous analysis would show that the distribution of shear varied markedly around the perimeter and was accompanied by torsional moments. Extensive experimental work shows that punching shear strength can be significantly reduced where substantial moment transfer occurs. This implies that punching shear is not an entirely plastic phenomenon, and the shear at the ultimate limit state cannot be fully redistributed around the perimeter. A way of dealing with this in design is to increase the design shear force by a factor which is a function of the geometry of the critical perimeter and the moment transferred. The provisions in EN 1992-1-1 introduce a multiplier, β, to increase the average shear stress around the perimeter such that

$$v_{Ed} = \beta V_{Ed}/u_i d \qquad\qquad (D6.14)$$

where v_{Ed} is the design ultimate shear stress on the perimeter considered, V_{Ed} is the design ultimate shear force on the perimeter considered, u_i is the length of the control perimeter considered, d is the mean effective depth of the slab at the control perimeter, and β is the shear multiplier, taking account of moment transfer.

The principle behind the definition of β is that the effect of transferring a moment between the slab and a column can be modelled by considering a distribution of shear around the control perimeter considered such that it provides a moment equal to the moment transferred. The distribution of shear assumed is sketched in Fig. 6.14a, which is redrawn after *Fig. 6.19* in EN 1992-1-1. It will be seen that the magnitude of the distributed shear force is a function of the moment transferred, the distance of the perimeter from the loaded area and the shape of the loaded area. It will be seen that

$$\beta V_{\mathrm{Ed}}/u_i = V_{\mathrm{Ed}}/u_i + \Delta v$$

hence

$$\beta = 1 + \Delta v/(V_{\mathrm{Ed}}/u_i) = 1 + \Delta v\, u_i/V_{\mathrm{Ed}} \qquad (D6.15)$$

In principle, Δv can be calculated from simple statics, since the moment transferred between the slab and the column, ΔM_{Ed}, must be equal to the moment produced by the shear Δv distributed around the perimeter, u_i. Inspection of Fig. 6.12b shows that this may be written as

$$\Delta M_{\mathrm{Ed}} = 4\,\Delta v[c_1/2 \times c_1/4 + c_2\,(c_1/2 + 2d) + 2d\pi/2(c_1/2 + 2d \times 2/\pi)]$$

$$= \Delta v(c_1^2/2 + c_1 c_2 + 4c_2 d + 16d^2 + 2\pi dc_1) = W_1\,\Delta v$$

Hence,

$$\Delta v = \Delta M_{\mathrm{Ed}}/W_1$$

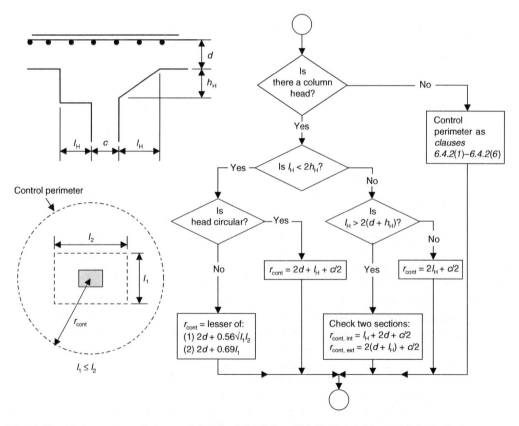

Fig. 6.13. Interpretation of *clauses 6.4.2(8)–6.4.2(11)* in EN 1992-1-1 for establishing the location of the basic control perimeter where column heads are used

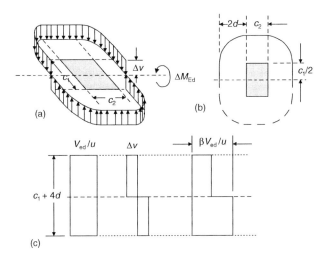

Fig. 6.14. Assumed shear distribution to resist moment transfer

Table 6.5. Values for the aspect ratio factor, k

c_1/c_2	k
≤ 0.5	0.45
1.0	0.60
2.0	0.70
≥ 3.0	0.80

Substituting for Δv into the equation for β gives

$$\beta = 1 + \Delta M_{\text{Ed}} u_i / W_1 V_{\text{Ed}}$$

It is found that a further correction factor is required to adjust the value of β for the aspect ratio of the column section. The full equation in the code is thus

$$\beta = 1 + k(\Delta M_{\text{Ed}}/V_{\text{Ed}})(u_i/W_1) \tag{D6.16}$$

The coefficient k is given as a function of the aspect ratio of the loaded area in *Table 6.1* in EN 1992-1-1 (repeated here for convenience as Table 6.5). In the table, c_1 is the dimension of the cross section perpendicular to the axis of bending, while c_2 is the dimension parallel to the axis of bending.

It should be noticed that $\Delta M_{\text{Ed}}/V_{\text{Ed}}$ is actually the eccentricity of the concentrated load relative to the centroid of the loaded area. In some equations, EN 1992-1-1 uses 'e' for this rather than $\Delta M_{\text{Ed}}/V_{\text{Ed}}$. It should also be noted that EN 1992-1-1 uses the term M_{Ed} for the moment transferred between the slab and the column. This seems a possible cause of confusion, since the moment transferred is not the moment in the slab but rather the difference in moment between one side of the column and the other. For this reason, ΔM_{Ed} has been used here for the moment transferred.

The derivation of the equation for β for a rectangular internal column has been derived in order to show the basic logic of the system. There are many other possible configurations of column, and EN 1992-1-1 occupies three pages setting out the appropriate values. For convenience, these are set out in Table 6.6.

For structures where the lateral stability does not depend on the frame action of the slabs and columns, EN 1992-1-1 permits the use of simplified values for β. These values are $\beta = 1.15$ for interior columns, $\beta = 1.4$ for edge columns and $\beta = 1.5$ for corner columns. These values are Nationally Determined Parameters.

Table 6.6. Values of the punching shear enhancement factor for various types of column

Case	Value for β
Internal rectangular column, uniaxial bending	$\beta = 1 + k(\Delta M_{Ed}/V_{Ed})(u_1/W_1)$ $W_1 = c_1^2/2 + c_1 c_2 + 4c_2 d + 16d^2 + 2\pi dc_1$ Values of k from Table 6.4
Internal rectangular column, biaxial bending	$1 + 1.8\sqrt{\{[\Delta M_{Edy}/(c_z + 4d)]^2 + [\Delta M_{Edz}/(c_y + 4d)]^2\}}$ ΔM_{Edy} and ΔM_{Edz} are respectively the moments transferred in the y and z directions while c_y and c_z are respectively the section dimensions in the y and z directions
Rectangular edge column; axis of bending parallel to the slab edge, eccentricity is towards the interior	$\beta = u_1/u_1^*$ (i.e. shear is assumed uniformly distributed over perimeter u_1^* as defined in *Fig. 6.20a* of EN 1992-1-1)
Rectangular edge column; bending about both axes. Eccentricity perpendicular to the slab edge is towards the exterior	$\beta = 1 + k(\Delta M_{Ed}/V_{Ed})(u_1/W_1)$ W_1 calculated by taking moments about the centroid of the control perimeter, u_1 Values of k from Table 6.4
Rectangular edge column; bending about both axes. Eccentricity perpendicular to the slab edge is towards the interior	$\beta = u_1/u_1^* + k(\Delta M_{Ed, par}/V_{Ed})(u_1/W_1)$ $W_1 = c_2^2/4 + c_1 c_2 + 4c_1 d + 8d^2 + \pi dc_2$ $\Delta M_{Ed, par}$ is the moment transfer about an axis perpendicular to the slab edge. k is determined from Table 6.4 with c_1/c_2 put equal to $0.5(c_1/c_2)$ c_1 is the section dimension perpendicular to the slab edge; c_2 is the dimension parallel to the slab edge
Rectangular corner column, eccentricity is towards the interior	$\beta = u_1/u_1^*$ (i.e. punching force is considered uniformly distributed along perimeter u_1^* in *Fig. 6.20b* of EN 1992-1-1)
Rectangular corner column, eccentricity is towards the exterior	$\beta = 1 + k(\Delta M_{Ed}/V_{Ed})(u_1/W_1)$ $W_1 = c_1^2/2 + c_1 c_2 + 4c_2 d + 16d^2 + 2\pi dc_1$ Values of k from Table 6.4
Interior circular column	$\beta = 1 + 0.6\pi(\Delta M_{Ed}/V_{Ed})/(D + 4d)$ D is the diameter of the column
Circular edge or corner columns	No information given

6.5.4. Punching shear resistance of slabs without shear reinforcement

Provided that $v_{Ed} \leq v_{Rd, c}$, no shear reinforcement is required. $v_{Rd, c}$ is given by

$$v_{Rd, c} = (0.18/\gamma_c)k(100\rho_1 f_{ck})^{1/3} + 0.1\sigma_{cp} \qquad \text{(D6.17)}$$

$v_{Rd, c}$ should not be taken as less than

$$0.035k^{3/2}f_{ck}^{1/2} + 0.1\sigma_{cp} \qquad \text{(D6.18)}$$

In equation (D6.17),

$$k = 1 + \sqrt{(200/d)} \leq 2.0$$

where d is in millimetres,

$$\rho_1 = \sqrt{(\rho_{1y}\rho_{1z})} \leq 0.02$$

where ρ_{1y} and ρ_{1z} are, respectively, the reinforcement ratios in the y and z directions calculated for the reinforcement within a width equal to the column dimension plus $3d$ on each side. σ_{cp} is the average longitudinal stress (note compression is positive).

These expresions only differ from those for beams unreinforced in shear by the necessity to obtain average values for ρ_1 and σ_{cp} and the reduction of the coefficient applied to the axial stress σ_{cp} from 0.15 to 0.1.

6.5.5. Reinforcement for punching shear

The punching strength of a slab with shear reinforcement is given for vertical shear reinforcement by

$$v_{\text{Rd, cs}} = 0.75v_{\text{Rd, c}} + 1.5(d/s_r)A_{\text{sw}}f_{\text{ywd, ef}}(1/u_1d) \tag{D6.19}$$

where A_{sw} is the area of one perimeter of shear reinforcement around the column, d is the average effective depth of the flexural reinforcement (in millimetres),

$$f_{\text{ywd, ef}} = 250 + 0.25\text{d} \le f_{\text{ywd}} \text{ N/mm}^2$$

(d in millimetres) and s_r is the radial spacing of perimeters of shear reinforcement (in millimetres)

It will be noted that the approach to the design of shear reinforcement is in principle different to that used in beams where the concrete is assumed not to contribute to the shear strength of a section reinforced in shear. Here, a contribution from the concrete is assumed but, unlike the general international consensus, the concrete contribution is reduced from the full strength of the section without shear reinforcement while cot θ is effectively assumed to be 1.5 rather than the more common 1.0. The formula has to be seen to be basically empirical, and one cannot avoid an impression that all the parameters have been changed slightly from previous formulae just for the sake of being different.

The concepts of a perimeter of shear reinforcement and the spacing of the perimeters are awkward, and will have to be interpreted for each particular case. A perimeter of reinforcement is presumably reinforcement provided on a perimeter parallel to the control perimeter u_1 but lying inside it. Unfortunately, it is not clear that shear reinforcement is actually provided in this way; indeed, *Fig. 6.22* of EN 1992-1-1 makes this difficulty clear. It is also unfortunate that the rules in *clauses 6.4.5(4)* and *9.4.4.3(1)* appear to be uninterpretable. The following is an attempt to develop a set of rules which can reasonably be interpreted in practice.

Clause 6.4.5(4)
Clause 9.4.4.3(1)

Having found that the shear reinforcement is required on the control perimeter at a distance of $2d$ from the face of the loaded area (first control perimeter), the control perimeter beyond which shear reinforcement is no longer required is found from

$$u_{\text{out, ef}} = \beta V_{\text{Ed}}/v_{\text{rd, c}}d \tag{D6.20}$$

having calculated $u_{\text{out, ef}}$, the distance of this perimeter from the loaded area may easily be calculated. While failure will not take place on this perimeter, it can occur on any perimeter within this (even, in theory, 1 mm inside it). The failure is predicted to occur over a distance, measured radially inwards from the perimeter, equal to $2d$. Shear reinforcement for this failure must, therefore, be within this zone between the perimeter and a parallel perimeter $2d$ inside it. Reinforcement close to either boundary of this zone is unlikely to be effective, and therefore a margin is needed on either edge. On the inner edge, *Fig. 9.12a* of EN 1992-1-1 indicates that reinforcement should not be closer than $0.3d$ to the loaded area, which constitutes the inner boundary of the failure zone for the first control perimeter. The maximum radial spacing of shear reinforcement is $0.75d$, and at least two sets of reinforcement are required. This would suggest, for the first control perimeter, that reinforcement could reasonably be provided at distances of $0.5d$ and $1.25d$ from the loaded area (column face), leaving a gap of $0.75d$ between the outermost reinforcement and the first control perimeter. It is suggested that this procedure is generalized into the following rules.

Considering any possible perimeter, u_n, lying between $u_{\text{out, ef}}$ and the first control perimeter, shear reinforcement should be arranged so that a total area of shear reinforcement equal to $A_{\text{sw, tot}}$ lies within a zone between a perimeter situated $0.3d$ and a perimeter situated $1.7d$ inside u_n. The radial spacing of the reinforcement should not exceed $0.75d$, and the

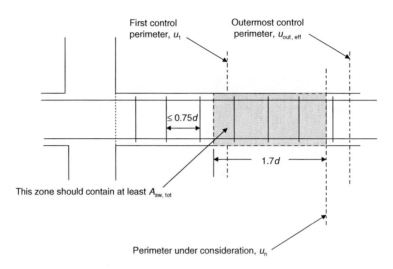

Fig. 6.15. Arrangement of punching reinforcement

circumferential spacing should not exceed $1.5d$ for the reinforcement for the first control perimeter or $2d$ for subsequent perimeters. $A_{sw,\,tot}$ is given by

$$A_{sw,\,tot} \geq u_n d(v_{Ed} - 0.75v_{Rd,\,c})/f_{ywd,\,eff}$$

This procedure is illustrated in Fig. 6.15.

Example 6.2: lightly loaded slab–column connection

Check the punching shear strength of the slab around an internal column supporting a 225 mm thick flat slab having 6 m spans in both directions. The column is 300×400 mm, and the design shear force established from analysis of the slab, which has 6.5 m spans in both directions and supports a design ultimate load of 9 kN/m², is 400 kN. Design of the slab for flexure gave an average value for the reinforcement ratio as 0.0077. The characteristic concrete strength is 30 N/mm². Bending is about the major axis of the column only, and the moment transferred between the slab and the columns is 45 kN m.

Assuming 20 mm cover and 12 mm diameter bars give the average effective depth as

$225 - 20 - 12 = 193$ mm

Table 4.8 of EN 1992-1-1 gives the basic design shear strength as 0.34 N/mm² for 30 N/mm² concrete. The critical perimeter is

$2(300 + 400) + 193 \times 4 \times 3.1416 = 3825$ mm

To find the design effective shear stress it is now necessary to calculate β. For an internal rectangular column, Table 6.6 gives

$$W_1 = c_1^2/2 + c_1 c_2 + 4c_2 d + 16d^2 + 2\pi dc_1$$

For the slab and column dimensions considered, this gives

$$W_1 = 400^2/2 + 400 \times 300 + 4 \times 300 \times 193 + 16 \times 193^2 + 2\pi \times 193 \times 400 = 1512.6 \times 10^3$$

$c_1/c_2 = 400/300 = 1.333$

From Table 6.5, $k = 0.63$. This gives

$$\beta = 1 + 0.63 \times 45 \times 10^6 \times 3825/(400 \times 1000 \times 1512.6 \times 1000) = 1.17$$

The effective design shear stress is

$$v_{Ed} = 400 \times 1000 \times 1.17/(3825 \times 193) = 0.633 \text{ N/mm}^2$$

The depth factor, k, for a slab with an effective depth less than 200 mm is 2.0. Equation (D6.17) gives

$$v_{Rd, c} = 0.18/1.5 \times 2.0 \times (100 \times 0.0077 \times 30)^{1/3} = 0.684$$

This exceeds 0.633 N/mm^2, hence no shear reinforcement is needed.

Example 6.3: heavily loaded slab–column connection requiring shear reinforcement

The slab–column connection considered in Example 6.2 will now be designed for an increased design shear force of 600 kN. All other factors will be assumed to be as in Example 6.2.

In calculating v_{Ed}, W_1 and c_1/c_2 will remain as in Example 6.2, but β will change because the eccentricity of the load will reduce. β can now be calculated as

$$\beta = 1 + 0.63 \times 45 \times 10^6 \times 3825/(600 \times 1000 \times 1512.6 \times 1000) = 1.12$$

v_{Ed} is now given by

$$v_{Ed} = 600 \times 1000 \times 1.12/(3825 \times 193) = 0.91 \text{ N/mm}^2$$

As before, $v_{Rd, c} = 0.684$, thus punching shear reinforcement is required.

It is convenient at this stage to calculate u_{outer} to establish the extent of punching reinforcement:

$$u_{outer} = 1.12 \times 600 \times 1000/(0.684 \times 193) = 5090 \text{ mm}$$

This can easily be calculated to be a perimeter at a distance

$$(5090 - 1400)/2\pi = 587 \text{ mm}$$

or $3.04d$ from the column face.

The area of reinforcement required across the failure zone can be calculated for the first control perimeter and for u_{outer}. At any point in between these, the required area can be found by linear interpolation. For the first control perimeter, the total area of reinforcement that must be provided within the perimeter is given by

$$A_{s, tot} = 3825 \times 193 \times (0.91 - 0.75 \times 0.684)/435 = 674 \text{ mm}^2$$

At the outer perimeter, the total area is given by

$$A_{s, tot} = 5090 \times 193 \times (0.91 - 0.75 \times 0.684)/435 = 896 \text{ mm}^2$$

Figure 6.16 shows the required steel area for perimeters within this area.

Assuming that the first perimeter of reinforcement is provided at 125 mm from the column face and then at 150 mm centres as far as necessary, the length of each perimeter can be calculated, and, hence, from the maximum spacing rules, the minimum number of bars which should be provided. For example, the length of a perimeter 125 mm from the column face is 2185 mm. The maximum permissible spacing around the perimeter is $1.5d = 290$ mm. To meet this requirement, $2185/290 = 8$ bars are required. Trial and error suggests that 8 mm diameter bars, which are the smallest bars generally available, should be used on all perimeters. Table 6.7 gives the steel areas supplied on each perimeter in this case.

The areas in Table 6.7 together with the assumption that reinforcement is only effective for a particular failure zone when it is at least $0.3d$ from the inner edge of the failure zone permits the total reinforcement provided for any outer perimeter of a failure zone to be calculated. This is plotted on Fig. 6.16. It will be seen that adequate reinforcement is provided at all perimeters and that all the detailing rules are obeyed. It will also be seen that the amount of reinforcement provided is actually defined by the minimum circumferential and radial bar spacing rules rather than the required strength. This seems likely to be generally the case. In particular, the radial maximum spacing of $0.75d$ ensures that, in many places, there are, unavoidably, three perimeters of reinforcement within the failure zone in many situations.

Table 6.7. Calculations for each perimeter of reinforcement in Example 6.3

	Perimeter of reinforcement		
	1	2	3
Distance from column face (mm)	125	250	375
(Distance from column face)/d	0.65d	1.3d	1.9d
Perimeter (mm)	2185	2970	3756
Maximum spacing	290	290	290
No. of bars	8	10	13
Bar diameter	8	8	8
Area on perimeter	402	603	553

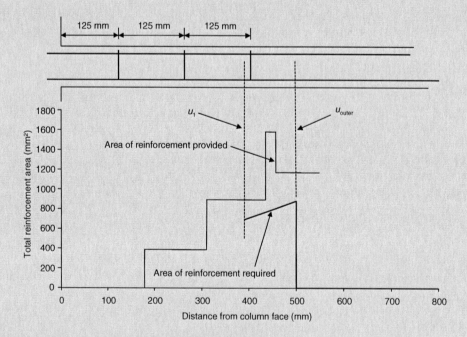

Fig. 6.16. Provision of shear reinforcement in Example 6.3

CHAPTER 7

Slender columns and beams

7.1. Scope

This chapter covers the material given in *clause 5.8* of EN 1992-1-1. Also covered is the material in *Annex D*, which provides supplementary information on global second-order effects.

Clause 5.8

7.2. Background to design of columns for slenderness effects

When an element is subjected to an axial load combined with a moment, it will deflect. This deflection will increase the moment at any section in the element by an amount equal to the axial force multiplied by the deflection at that point. This extra moment will cause the resistance of the element to be reduced below that calculated ignoring the deflections. In many, if not most, practical situations, the effect of deflections is so small that it can be ignored. Indeed, in many practical cases the effect of deflections on the design of a member is not merely insignificant but actually nil. How this comes about will be explained later. A slender member is one where the influence of deflections on the ultimate strength is significant and must be taken into account in the design.

Slenderness effects in struts have been the subject of theoretical and experimental study from the earliest days of the science of structural mechanics. Euler (1707–1783) seems to have been the first person to study this problem mathematically, and the Euler equation for the buckling strength of a strut must be well known to any engineer. The Euler equation deals with the failure by instability of a strut made of infinitely strong, elastic material. Failure occurs not due to failure of the material but because the deflection under the critical buckling load becomes infinite. The problem that the designer faces in the design of reinforced concrete members is very rarely one of classical buckling; failure of the material occurs at a lower load than would be predicted if the deflections were ignored, due to the moments being somewhat higher than predicted. The other critical difference from classical buckling theory is that reinforced concrete is definitely not elastic at the ultimate limit state.

In order to understand what the design rules given in EN 1992-1-1 are aiming to achieve, it is necessary to have an idea of the forms of deflection which may occur. There are two modes of deflection which are normally possible: sidesway of the whole structure, and deflection of a single column without sidesway. These two possibilities are shown schematically in Fig. 7.1. It will be seen that the two modes give very different forms of bending moment diagram in the columns. In the sidesway mode it is important to note that all the columns within the storey are subjected to the same deflection.

Code rules dealing with slenderness effects commonly include three basic steps:

(1) establish which modes of deflection are likely
(2) establish whether slenderness effects are likely to be significant (i.e. whether the structure or member considered is slender)
(3) if the slenderness effects are likely to be significant, design to make allowance for them.

This is the approach adopted by EN 1992-1-1, and the basic principles will be introduced here. Detailed guidance with use of the provisions will be given later.

In order to establish the deflection modes which should be considered, EN 1992-1-1 uses the following classification system. Structures are classified as:

• sway or non-sway and
• braced or unbraced.

A sway structure is one where sidesway (global second-order effects) is likely to be significant (as in Fig. 7.1a), while in non-sway structures, sidesway is unlikely to be significant. 'Significance' is defined in this context as a situation where lateral displacement of the ends of the columns increase the critical bending moments by more than 10% above the moment calculated ignoring the displacements. Basically, whether a structure is sway or non-sway will depend upon its stiffness against lateral deformation. In most cases a structure can be classified by inspection. Section 7.3.6 of this guide deals with this question in more detail.

A braced structure is one which contains bracing elements. These are vertical elements (usually walls) which are so stiff relative to the other vertical elements that they may be assumed to attract all horizontal forces. In general, where the bracing elements in a braced structure consist of shear walls or cores, braced structures may be assumed to be non-sway without further checking.

Having classified the structure, the following can be said:

• Non-sway structures: only the deformation of individual columns of the type illustrated in Fig. 7.1b need be considered as, by definition, sidesway will be insignificant.
• Sway structures:
   – Braced structures. As mentioned above, braced structures may normally be assumed to be non-sway unless the bracing elements are relatively flexible. If they are, then the bracing elements or structure should be analysed for the effects of sidesway but

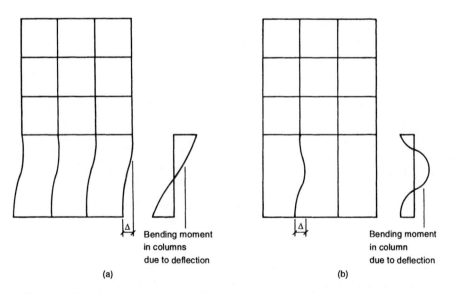

(a)   (b)

Fig. 7.1. Modes of deflection of columns in a structure: (a) sidesway of whole structure; (b) deflection of a single column

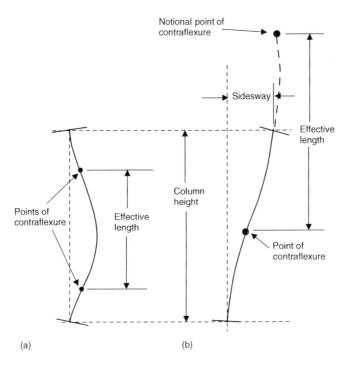

Fig. 7.2. The concept of effective length: (a) isolated column; (b) column with sidesway

the braced elements within the structure may be assumed to deform, as illustrated in Fig. 7.1b, and sidesway may be ignored.
- Unbraced structures. These structures should first be designed to take account of sidesway of the whole structure as illustrated in Fig. 7.1a, and then each column in turn should be analysed for the possibility of its deforming, as illustrated in Fig. 7.1b.

The second element in the code provisions is establishing whether slenderness effects are likely to be significant or whether they can be ignored. EN 1992-1-1 defines significance as being when slenderness effects increase the first-order moments by more than 10%. It is useful to have this stated as a general principle, but, from the practical point of view, it is unhelpful since it can only be established that the effect is less than 10% by doing a full calculation while the objective is to establish when the calculation is unnecessary. Some simplified rules which may be deemed to satisfy this requirement are therefore needed. As with most other codes, EN 1992-1-1 formulates these rules in terms of two parameters:

- the effective length of the column
- the slenderness ratio.

Effective length is a concept used almost universally in the treatment of second-order effects. The effective length can best be defined as the length of a pinned ended strut having the same cross-section as the column considered and which would have the same buckling strength. The concept is illustrated in Fig. 7.2. The effective length will be seen to depend upon the type of deflection considered and the end fixity of the column. A non-sway column will have an effective length which is between 0.5 and 1.0 times the actual clear height of the column while a sway column will have an effective length greater than the clear height. Only in very rare cases will the effective height exceed twice the overall height. Formulae can be derived which will give the effective height as a function of the stiffnesses of the column and the members framing into each end of the column. The following equations are given in EN 1992-1-1 to establish the effective length:

(1) For isolated columns in non-sway frames (braced columns):

Table 7.1. Approximate values for effective length factors (effective length = factor × clear height)

End condition at top	End condition at bottom		
	1	2	3
Braced or non-sway columns			
1	0.75	0.80	0.90
2	0.80	0.85	0.95
3	0.90	0.95	1.00
Sway columns			
1	1.2	1.3	1.6
2	1.3	1.5	1.8
3	1.6	1.8	–
4	2.2	–	–

$$l_0 = 0.5l \sqrt{\left(1 + \frac{k_1}{0.45 + k_1}\right)\left(1 + \frac{k_2}{0.45 + k_2}\right)} \qquad \text{(D7.1)}$$

(2) For columns subjected to sway:

$l_0 = 1 \times$ the greater of a or b below:

$$a = \sqrt{1 + 10\frac{k_1 k_2}{k_1 + k_2}} \qquad \text{(D7.2)}$$

$$b = \left(1 + \frac{k_1}{1 + k_1}\right)\left(1 + \frac{k_2}{1 + k_2}\right) \qquad \text{(D7.3)}$$

where l_0 is the effective length of the column, l is the clear height of the column between restraints, k_1 is the ratio of the column stiffness to the stiffness of the upper joint and k_2 is the ratio of the column stiffness to the stiffness of the lower joint

These equations are time-consuming to use, and Table 7.1 (from BS 8110) provides a simple and conservative means of estimating the effective length for columns in regular structures. The values in the table are difficult to justify rigorously, and the interpretation of the clauses in EN 1992-1-1 are, in any case, unclear. The lack of clarity results from the definition of k_1 and k_2. k, which may be either k_1 or k_2, is defined by

$$k = (\theta/M)(EI/l) \qquad \text{(D7.4)}$$

where θ is the rotation of the joint at the end of the column considered for bending moment M applied to the restraining members by the buckling member or members. EI/l is the bending stiffness of the compression member or members considered to be buckling.

If there is a compression member above the joint at the upper end of the column considered or below the joint at the lower end of the column considered then it is necessary to consider whether the column is likely to contribute to the deflection or to restrain it. If the column is likely to contribute to the deflection, then EI/l should be replaced by $(EI/l)a + (EI/l)b$, where a and b indicate the columns above and below the joint. If, on the other hand, the column is likely to act to restrain the deflection then, logically, its stiffness should be included in the calculation of θ/M, though this is not stated in the code. In general, it is believed that the possibility of the columns above and below a joint buckling at the same time is improbable and, though the column will deflect somewhat, its deflection will be much smaller than that of the failing column. The values in Table 7.1 are roughly consistent with assuming that the outer columns neither drive nor restrain the deflection, and this seems to give a practically reasonable value for effective length.

The definitions of the four end conditions in Table 7.1 are as follows:

- **Condition 1.** The end of the column is connected monolithically to beams on either side which are at least as deep as the overall dimension of the column in the plane considered. Where the column is connected to a foundation, this should be of a form specifically designed to carry moment.
- **Condition 2.** The end of the column is connected monolithically to beams or slabs on either side which are shallower than the overall dimension of the column in the plane considered.
- **Condition 3.** The end of the column is connected to members which, while not specifically designed to provide restraint to rotation of the end of the column, will, nevertheless, provide some nominal restraint.
- **Condition 4.** The end of the column is unrestrained against both lateral movement and rotation (e.g. the free end of a cantilever column in an unbraced frame).

The other factor used to define the slenderness of a column is the slenderness ratio. This is defined as the ratio of the effective length to the radius of gyration. The radius of gyration of a section is defined by

$$i = I/A$$

where i is the radius of gyration, I is the second moment of area of the cross-section and A is the area of the cross-section. For a rectangular section, $I = bh^3/12$ and $A = bh$, hence the radius of gyration $= 0.2887h$, where h is the dimension of the section perpendicular to the axis of bending considered. The radius of gyration of a circular section is equal to its actual radius.

Many codes define the slenderness ratio as the ratio of the effective length to the section depth. It is not easy to argue that one method is intrinsically better than another for reinforced concrete since, as discussed earlier, the problem being dealt with is not one of classical buckling and nor is reinforced concrete an elastic material. Use of the radius of gyration does, however, have the advantage of providing a reasonable way of dealing with non-rectangular sections.

EN 1992-1-1 gives limits for the slenderness ratio above which it will be necessary to take account of slenderness effects for non-sway columns. If the slenderness ratio is less than λ_{lim}, then second-order effects may be ignored.

$$\lambda_{\text{lim}} = 20ABC/\sqrt{n} \tag{D7.5}$$

where

$$n = N_{\text{Ed}}/A_{\text{c}}f_{\text{cd}} \tag{D7.6}$$

and A, B and C are Nationally Defined Parameters, and their recommended values are as follows.

$$A = 1/(1 + 0.2\varphi_{\text{ef}}) \tag{D7.7}$$

where φ_{ef} is the effective creep coefficient. Details of how to assess φ_{ef} are given in *clause 5.8.4* of EN 1992-1-1, but, in the absence of information, A may be taken as 0.7.

Clause 5.8.4

$$B = \sqrt{(1 + 2A_{\text{s}}f_{\text{yd}}/Af_{\text{cd}})} \tag{D7.8}$$

Since the objective of the calculation is usually to find the value of A_{s}, this is unlikely to be known, though a lower-bound estimate may be made by finding $A_{\text{s}}f_{\text{yd}}/Af_{\text{cd}}$ for the first-order moments using the appropriate design chart from Chapter 5. Alternatively, a value of 1.1 may be assumed.

$$C = 1.7 - M_{01}/M_{02} \tag{D7.9}$$

where M_{01} and M_{02} are the end moments of the column chosen so that $|M_{02}| \geq |M_{01}|$. In the expression for C, M_{02} and M_{01} should be given their correct relative algebraic signs, thus, for the normal case of a column in double curvature, C will always be greater than 1.7.

On this basis, for a column in double curvature, a conservative minimum value for λ_{lim} would be $27/\sqrt{n}$.

No simple limits are given in EN 1992-1-1 for sway frames, but it seems reasonable to assume that any structure that is classifiable as a sway frame will need to be designed for the effects of deformations.

7.3. Design for slenderness effects

7.3.1. Basic approaches

Having concluded that a column or structure requires to be designed for slenderness effects, the remaining, and biggest, problem is how this should be done. The rest of this chapter will be concerned with this problem.

EN 1992-1-1 provides four basic approaches:

(1) A general method based on non-linear analysis of the structure including geometric non-linearity. This approach will require a suitable computer package, and will not be considered further here. For a relatively rigorous analysis of this type, it is necessary to know the reinforcement details before the analysis is attempted, and so it can only be used as a check or iteratively.

(2) A second-order analysis based on the nominal stiffnesses of the column or structure. This can be done using less sophisticated software than the first approach, but still requires knowledge of the reinforcement areas. It can therefore also only be used iteratively or as a check. Establishment of the nominal stiffness is set out in *clause 5.8.7.2* of EN 1992-1-1. Since this is required for the moment magnification method discussed below, this will be considered further when the moment magnification method is discussed.

Clause 5.8.7.2

(3) The moment magnification factor method. In this method, the design moment including second-order effects is obtained by multiplying the first-order moment by a factor. This method will be discussed in more detail below.

(4) The nominal curvature method. In this method, an estimate is made of the ultimate deflection, and, from this, an estimate of the second-order moment which is added to the first-order moment. This is in principle the method currently used in BS 8110, and should be familiar to UK designers. The method will be described in more detail below.

7.3.2. First-order moments

Both the moment magnification method and the nominal curvature method require an estimate of the first-order moment at around mid-height of the column, since this is the point where the second-order moment will be a maximum. Both methods use the same method to establish the first-order moment, which is as follows.

Classic analyses of buckling commonly considers the deformation of a pinned ended strut, but this is not the normal configuration of a column in a building. A normal column built monolithically into a structure at its top and bottom will deform, and be subjected to moments like those shown schematically in Fig. 7.3. It will be seen that the section of the column between the points of contraflexure in the final state of the column may be considered to be a pinned ended strut equivalent to that for which the analysis was carried out. The distance between the points of contraflexure is the effective length of the column. The estimation of this length has been covered earlier. The maximum moment due to deflection will be seen to occur at mid-height of the effective column. This will normally be somewhere close to mid-height of the real column. Clearly, the total moment to which the critical section is subjected is made up of the maximum moment due to the deflection plus the first-order moment at this height plus any allowance for accidental effects. A reasonable estimate of the first-order moment near mid-height of the column is given by

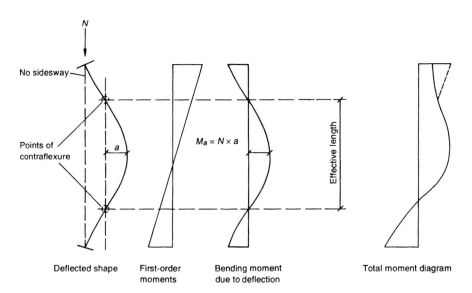

Fig. 7.3. Moments and deformations of a braced isolated column

$$M_{0e} = 0.6M_{max} + 0.4M_{min} \qquad\qquad (D7.10)$$

where M_{max} and M_{min} are, respectively, the numerically greater and lesser end moments. This relationship, written in terms of eccentricities, is included in EN 1992-1-1 (*clause 5.8.8.2*) together with a further precautionary limitation that the moment should not be less than $0.4M_{max}$.

Clause 5.8.8.2

All methods require that the possibility of the structure being accidentally constructed out-of-plumb should be considered. This is allowed for by introducing an accidental eccentricity of the axial load in the case of an isolated column or by considering an accidental inclination from the vertical for a sway frame. *Clause 5.2* defines the accidental inclination as an angle θ_i in radians:

Clause 5.2

$$\theta_i = \alpha_h \alpha_m / 200 \qquad\qquad (D7.11)$$

where α_h is a reduction factor for height ($= 2/3 \leq 2/\sqrt{L} \leq 1$), α_m is a reduction factor for the number of members ($= \sqrt{[0.5(1 + 1/m)]}$), L is the length of the member or height of the building and m is the number of vertical members contributing to the total effect.

For an isolated member such as a braced column, L is the actual length of the column, and $m = 1$. In this circumstance the effect can be considered to be given by an additional eccentricity at mid-height equal to $\theta_i l_0/2$. This may always be conservatively assumed to be equal to $l_0/400$, resulting in a moment of

$$M_i = N_{Ed} l_0 / 400$$

From *Fig. 5.1a2* in EN 1992-1-1, it is clear that this accidental moment is a maximum at mid-height of the column. It follows that the accidental moment should not be added to M_{max} and/or M_{min} but should be added to M_{0e}. Thus,

$$M_{0Ed} = M_{0e} + M_i$$

7.3.3. Moment magnifier method

At first glance, this method seems deceptively simple. More detailed investigation shows that it is actually quite complex to apply. To aid application, a flow chart has been provided (Fig. 7.4) – this has been slightly simplified from the full description as it has been drawn specifically for columns in double curvature. That is, columns where the moment at one end is either zero or has the opposite sign to that at the other end. The notation used in Fig. 7.4 is that used in the code, but, for convenience, is repeated below:

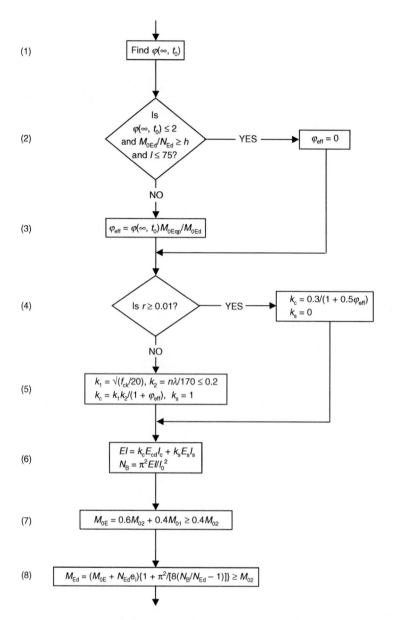

(1) Find $\varphi(\infty, t_0)$

(2) Is $\varphi(\infty, t_0) \leq 2$ and $M_{0Ed}/N_{Ed} \geq h$ and $l \leq 75$? — YES → $\varphi_{eff} = 0$

NO

(3) $\varphi_{eff} = \varphi(\infty, t_0)M_{0Eqp}/M_{0Ed}$

(4) Is $r \geq 0.01$? — YES → $k_c = 0.3/(1 + 0.5\varphi_{eff})$, $k_s = 0$

NO

(5) $k_1 = \sqrt{(f_{ck}/20)}$, $k_2 = n\lambda/170 \leq 0.2$, $k_c = k_1 k_2/(1 + \varphi_{eff})$, $k_s = 1$

(6) $EI = k_c E_{cd} I_c + k_s E_s I_s$, $N_B = \pi^2 EI/l_0^2$

(7) $M_{0E} = 0.6M_{02} + 0.4M_{01} \geq 0.4M_{02}$

(8) $M_{Ed} = (M_{0E} + N_{Ed}e_i)\{1 + \pi^2/[8(N_B/N_{Ed} - 1)]\} \geq M_{02}$

Fig. 7.4. Moment magnifier method flow chart

A_c	area of the concrete section
A_s	total area of the reinforcement
E	nominal modulus of elasticity of the column
E_{cd}	design modulus of elasticity of the concrete
E_s	modulus of elasticity of the reinforcement
I	nominal second moment of the area of the section about the axis of bending considered
I_c	second moment of the area of the concrete cross-section
I_s	second moment of the area of reinforcement about the centroid of the concrete section
k_1	concrete strength factor
k_2	axial force and slenderness factor
k_c	creep effect factor
k_s	reinforcement contribution factor

l_0 effective length of the column
M_{01} numerically smaller end moment from first-order analysis
M_{02} numerically larger end moment from first-order analysis
M_{0Ed} design first-order moment at about mid-height of the column
M_{0Eqp} design first-order moment under the quasi-permanent load combination (also at about mid-height of the column)
M_{Ed} design ultimate moment
N_B buckling load based on the nominal stiffness $(= \pi^2 EI/l_0^2)$
N_{Ed} design ultimate axial force
n relative axial force $(= N_{Ed}/A_c f_{cd})$
$\varphi(\infty, t_0)$ final creep coefficient of concrete loaded at time t_0
φ_{eff} effective creep coefficient
λ slenderness ratio
ρ reinforcement ratio $(= A_s/A_c)$

The various steps in the flow chart are numbered 1 to 8, and notes are given on each stage below:

(1) The creep coefficient can be obtained from *clause 3.1.4* in EN 1992-1-1.   *Clause 3.1.4*
(2) This provides a simplification for some situations where creep is unlikely to have any significant effect on the behaviour.
(3) The assumption is made that creep dominantly occurs under the quasi-permanent load and that the effect on creep of the additional load above the quasi-permanent load is relatively short-term. It may be noted that, in internal columns in frames with uniform spans, the quasi-permanent load is unlikely to produce significant moment, and hence M_{0Eqp} will be close to zero, resulting in φ_{eff} close to zero.
(4) This permits a simplification in the calculation of the effective stiffness where the reinforcement percentage exceeds 1%. A reasonable approach may be to assume that the reinforcement percentage will exceed 1% and use the simplified route generally, with a check at the end of the design procedure that this is actually the case.
(5) k_1 and k_2 indirectly take account of cracking as well as other factors.
(6) EI is the effective stiffness of the column section.
(7) This formula is identical to that in BS 8110. It aims to give an estimate of the moment around mid-height of the column where the additional moment due to deflection will be a maximum, $|M_{02}| \geq |M_{01}|$. In evaluating the formula, the moments should be given their correct algebraic sign.
(8) The factor 8 in the denominator of the expression is correct for columns in double curvature. Different values are given in *clause 5.8.7.3* of EN 1992-1-1 for other shapes of   *Clause 5.8.7.3* bending moment diagram. The final requirement that M_{Ed} should not be less than the numerically larger first-order moment is not included in the code, but it would seem nonsense if the design ultimate moment for a slender column was smaller than the moment that would be used if the column was short. Another way of considering this is that if the column was, say, stiffer than had been assumed and the deflection did not occur, then it would be the moment at the end which would be critical and not the moment near mid-height.

The difficulty with this method of establishing the moments in a slender column is that, though considerable efforts have gone into calibrating the method to give similar results as other methods, it provides no clear picture of the actual behaviour. In this respect, the nominal curvature method is considerably superior.

7.3.4. Nominal curvature method
This method is set out in EN 1992-1-1 in *clause 5.8.8*.   *Clause 5.8.8*
    If a rigorous non-linear analysis is carried out for a slender column, it will result in a load–deflection curve of a form similar to one of the curves in Fig. 7.5, depending on the

slenderness ratio. The object of a rigorous analysis is to calculate such curves, and hence find the maximum load capacity. The ideal simplified method would aim to establish this ultimate load, or something close to it, in a single calculation without the necessity to calculate the full load–deflection response. What the method in EN 1992-1-1 aims to do is to predict the deflection at which failure of the concrete will initiate (i.e. when the maximum compressive strain = ε_u). It will be seen from Fig. 7.5 that, if this point can be established, then it will either correspond to the actual ultimate load or to a conservative estimate of the ultimate load. Such a method will therefore, in principle, give a lower bound estimate of the strength. The calculation of the ultimate deflection is carried out as follows for a pinned ended strut:

(1) **Estimate the ultimate curvature of the section**. For a balanced section, the strain at the compression face of the section is 0.0035, while the strain in the reinforcement is given by f_{yd}/E_s. From this, the curvature can be written as

$$1/r_0 = (f_{yd}/E_s)/0.45d \tag{D7.12}$$

For axial loads above the balance point, the curvature will be less than this, as can be seen from Fig. 7.6. To allow for this, the balanced curvature is multiplied by a factor k_r, where

$$k_r = (n_u - n)/(n_u - n_{bal}) \leq 1$$
$$n_u = 1 + A_s f_{yd}/A_c f_{cd} \tag{D7.13}$$

and n_{bal} is the design load capacity of a balanced section. For a symmetrically reinforced rectangular section, this may be taken as 0.4.

It will be seen that, for a balanced section, $k_r = 1$, and that k_r reduces to zero as the load approaches the axial load capacity of the section. This approach provides an approximate estimate of the ultimate curvature. A better method might be to define k_r as ε_u/x, though this may not be so practically convenient.

Figures 7.7 and 7.8 can be used to obtain values of k_r for symmetrically reinforced rectangular sections.

The curvature at the critical section may also be increased by creep. This is allowed for by multiplying the curvature by a factor k_φ, where

$$k_\varphi = 1 + (0.35 + f_{ck}/200 - \lambda/150)\varphi_{ef} \tag{D7.14}$$

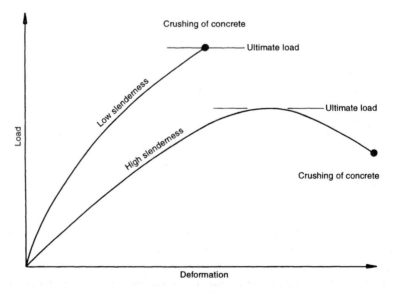

Fig. 7.5. Load–deformation curves for columns of different slenderness

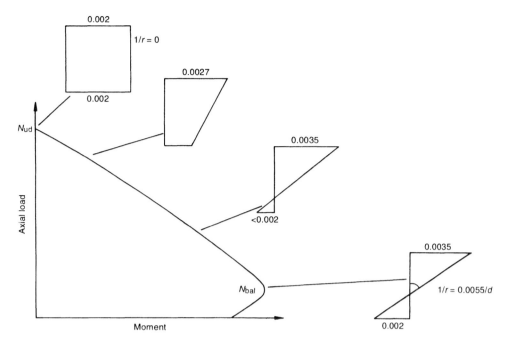

Fig. 7.6. Variation in curvature in a column at different levels of axial load

(2) **Calculate the ultimate deflection.** This may be done by assuming that the column deflects in the form of a sine curve. Thus, if it is assumed that the curvature is proportional to the moment, which is equal to the deflection times the axial load, we can write

$$a = \iint (1/r) \sin(\pi x / l_0) \, dx$$

$$= l_0^2 (1/r) \sin(\pi x / l_0) / \pi^2$$

at mid-height, $x/l_0 = 0.5$ and, hence, approximately,

$$a = 0.1(1/r) l_0^{\,2}$$

The ultimate deflection at mid-height of the column can thus be calculated. On the one hand, the estimate is likely to be conservative since the assumption of a sinusoidal variation of curvature over the height of the column will be conservative due to the non-linear behaviour of reinforced concrete. On the other hand, some of this conservatism will be counterbalanced by the coefficient k_r giving an unconservative estimate of the curvature in some cases.

In EN 1992-1-1, the deflection is, for convenience, given as an eccentricity, and, thus,

$$e_2 = 0.1(1/r) l_0^{\,2} \tag{D7.15}$$

where e_2 is the second-order eccentricity.

(3) **Define the design moments in the structure.** This is covered in Section 7.3.2. The design moment is now given by

$$M_{\mathrm{Ed}} = M_{0\mathrm{Ed}} + M_2$$

where $M_2 = N_{\mathrm{Ed}} e_2$.

It will be seen from Fig. 7.3 that as well as the moment near mid-height being modified by the deflections, the end moments are also affected. The numerically larger end moment will actually be reduced while the numerically smaller end moment will be increased. Thus, rigorously, there are three possible conditions which should be considered in establishing the critical conditions for section design:

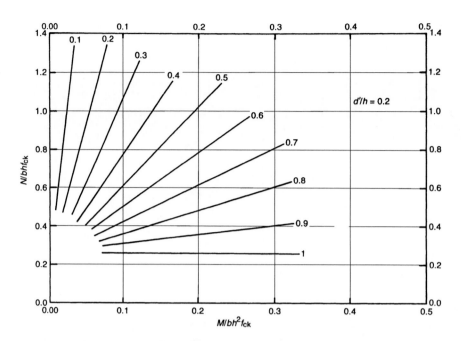

Fig. 7.7. k_2 values for section with $d'/h = 0.2$

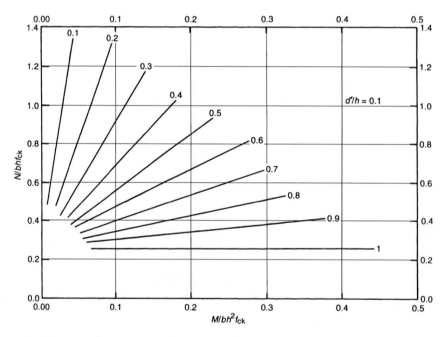

Fig. 7.8. k_2 values for section with $d'/h = 0.1$

(1) The numerically larger end moment assuming that no deflection occurs. In most columns of intermediate slenderness, this is the critical design condition and,though the deflection influences the moments at mid-height of the column, it does not influence the design. In EN 1992-1-1 the range of columns where this can be relied upon to be the case are those which have slenderness ratios lying below λ_{crit}. Nevertheless, columns may frequently occur where the design eccentricity given by *equation* (*5.31*) in EN 1992-1-1 is less than the larger end eccentricity. Clearly, the larger end eccentricity must be taken as the design value in this case.

(2) The moment at mid-height, allowing for the ultimate deflection. This is the condition covered explicitly in EN 1992-1-1.

(3) The numerically smaller end moment increased by the effects of deflections. The increase will be less than that corresponding to the second-order eccentricity at mid-height. The UK code, BS 8110, assumes that the second-order moment at this end of the column is equal to half the second-order moment at mid-height. EN 1992-1-1 does not require a check at this point, and, in fact, it is rarely critical.

Example 7.1: uniaxially bent rectangular column

Calculate the design ultimate moment for which the column shown in Fig. 7.9 should be designed. The column is in a braced frame. The stiffness of the beams and columns framing into the column considered may all be assumed to have the same value of L/EI as the column. It may be assumed that the column is loaded at 30 days, giving an effective creep coefficient $\varphi_{\text{eff}} = 0.87$.

(1) Calculate the effective height:
From *equation (5.15)* of EN 1992-1-1, $K_1 = K_2 = 0.75$, and hence the effective length is $0.81 \times 7.0 = 5.67$ m.

(2) Calculate the slenderness ratio:

slenderness ratio = effective height/radius of gyration.

$$= 5670/(0.2887 \times 300) = 65.5$$

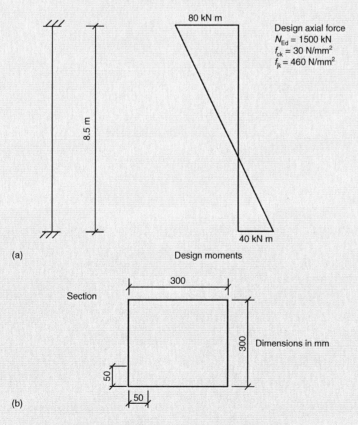

80 kN m

Design axial force
$N_{\text{Ed}} = 1500$ kN
$f_{\text{ck}} = 30$ N/mm²
$f_{\text{jk}} = 460$ N/mm²

8.5 m

40 kN m

(a)

Design moments

Section

300

300

Dimensions in mm

50

50

(b)

Fig. 7.9. Details of the column in Example 7.1

(3) Calculate the slenderness limits:
Equation (5.13N) in EN 1992-1-1 gives

$$\lambda_{min} = 20ABC/\sqrt{n}$$

$$A = 1/(1 + 0.2\varphi_{eff}) = 1/(1 + 0.2 \times 0.87) = 0.85$$

Since the reinforcement is not known, B will have to be assumed as suggested in the code as 1.1.

$$r_m = -40/80 = -0.5$$

giving $C = 2.2$.

$$n = 1500 \times 1000/(300^2 \times 20) = 0.83$$

and hence

$$\lambda_{min} = 20ABC/\sqrt{n} = 20 \times 0.85 \times 1.1 \times 2.2/\sqrt{0.83} = 45.15$$

The column is thus slender, and, since the slenderness exceeds λ_{lim}, specific measures to deal with slenderness are necessary.

(4) Calculate M_{0Ed}:
M_{0e} is the greater of

$$0.6 \times 80 - 0.4 \times 50 = 28$$

or

$$0.4 \times 80 = 32$$

and hence $M_{0e} = 32$. The accidental eccentricity is

$$e_i = l_0/400 = 5670/400 = 14.2$$

and hence the accidental moment is

$$1500 \times 14.2/1000 \text{ kN m} = 21.3 \text{ kN m}$$

The total first-order design moment is thus

$$32 + 21.3 = 53.3 \text{ kN m}$$

The design will now be carried out using the nominal curvature method.

(5) Calculate the second-order eccentricity, e_2:
The ultimate eccentricity at about mid-height is given as

$$0.1l_o^2K_rK_\varphi(\varepsilon_{yd}/0.45d)$$

From *equation (5.37)* in EN 1992-1-1 it will be found that $K_\varphi = 1.06$. For 500 N/mm^2 reinforcement,

$$\varepsilon_{yd} = 500/(1.15 \times 200\,000) = 0.00217$$

Since, from Fig. 7.9, it can be seen that

$$d = 300 - 50 = 250 \text{ mm}$$

the eccentricity is thus given by

$$0.00217 \times 5670^2 \times K_r \times 1.06/(0.45 \times 250 \times 10) = 65.8K_r \text{ mm}$$

This gives the second-order moment as

$$65.8 \times 1500 \times K_r/1000 \text{ kN m} = 98.7K_r \text{ kN m}$$

A value for K_r can be found from Fig. 7.7 by iteration in the following way:

$N/bhf_{ck} = 1500 \times 1000/(300 \times 300 \times 30) = 0.56$

A first estimate of the moment may be obtained by assuming $K_r = 1$. This gives

$M/bh^2f_{ck} = (53.5 + 98.7) \times 10^6/(300^3 \times 30) = 0.19$

From Fig. 7.7, this gives $K_r = 0.73$. This leads to a modified total moment of

$M/bh^2f_{ck} = (53.5 + 0.73 \times 98.7) \times 10^6/(300^3 \times 30) = 0.155$

From Fig. 7.7, this leads to a second estimate of K_r of 0.69 and a modified value of $M/bh^2f_{ck} = 0.15$. This is close enough to the previous result to suggest that this is a reasonably accurate assessment of K_r. Figure 5.9 may now be used to obtain a value for $\rho f_{yk}/f_{ck}$ of 0.51, and hence a steel area of 2754 mm^2.

The calculation may now be repeated using the moment magnification method.

(6) Calculate *EI* and the buckling load, N_B:

$EI = K_1K_2E_{cd}I_c/(1 + \varphi_{eff}) + E_sI_s$

$K_1 = \sqrt{(f_{ck}/20)} = \sqrt{(30/20)} = 1.22$

$K_2 = n\lambda/170 = 0.56 \times 65.5/170 = 0.216$

$E_{cd} = E_{cm}/1.2 = 32\,000/1.2 = 26\,700$ N/mm^2

$I_c = bh^3/12 = 300^4/12 = 675 \times 10^6$ mm^4

Using the steel area calculated by the nominal curvature method, I_s may be calculated as

$1 \times 2754 \times 100^2/2 = 27.5 \times 10^6$ mm^4

The buckling load may now be calculated:

$N_B = \pi^2 EI/l_0^2$

$\quad = \pi^2[1.22 \times 0.216 \times 26700 \times 675 \times 10^6/(1 + 0.87) + 200\,000 \times 27.5 \times 10^6]/(5670^2 \times 1000)$

$\quad = 2468$ kN

(7) Calculate the design moment:
The design moment is now given by

$M_{Ed} = M_{0Ed}[1 + \beta/(N_B/N_{Ed} - 1)]$

where $\beta = \pi^2/8$. Thus,

$M_{Ed} = 53.3[1 + (\pi^2/8)/(2468/1500 - 1)] = 155$ kN m

This compares with 126 kN m obtained using the nominal curvature method.

7.3.5. Other factors

The sections above have been concerned exclusively with cases where the second-order effects act about the same axis as the first-order moments and where only uniaxial moments act. While this will frequently be the case, it will by no means always be so, and *clause 5.8.2.P(4)* of EN 1992-1-1 states that: '*the structural behaviour shall be considered in the direction in which deformations can occur, and biaxial bending shall be taken into account when necessary*'. This clause is difficult to interpret, and much less clear than the equivalent requirements of either BS 8110 or the pre-standard ENV 1991-1.

Clause 5.8.2.P(4)

In principle, it seems possible that the accidental eccentricity, e_i and the second-order eccentricity, e_2, could occur about either axis of the column, independently of the first-order moments. This certainly must be true of the accidental eccentricity. It is also obviously true that where a column with one dimension much greater than the other was designed for first-order bending about the major axis, then it is highly likely that the second-order eccentricity will occur perpendicularly to the minor axis. How, then, should one decide upon a reasonable design situation to choose for the general case? The following proposal is not given in any code, but seems a reasonable way to reduce the problem to consideration of a single load case.

The column will deflect under the action of the first-order moments and the accidental moment. It is proposed that the second-order moment will occur in whichever direction the deflection due to the first-order moment as a proportion of the effective length of the column (a/l_0) is greatest. It is assumed, though this is not stated in the code, that the accidental moment and the second-order moments will only occur in one direction and not in both directions at once.

An approximate analysis shows that the ratio a/l_0 is proportional to $M\lambda/bh^2$, where h is the dimension of the cross-section perpendicular to the axis of bending and b is the dimension parallel to the axis of bending. It now remains to establish what situation gives the maximum value for this parameter and about which axis of bending. In the following, the notation shown in Fig. 7.10 is used. It is unimportant which is the minor and which is the major axis of the section. We will now define η_z and η_y as follows:

$$\eta_z = (M_{0ez} + N_{Ed}e_{iy})\lambda_z/h$$

$$\eta_y = (M_{0ey} + N_{Ed}e_{iz})\lambda_y/b$$

The subscripts y and z refer to moments, eccentricities and slenderness for bending about the y and z axes, respectively.

It is assumed that if $\eta_z \geq \eta_y$ then the second-order deflection will occur about the z axis, otherwise it will occur about the y axis. On this basis, the design moments about the z and y axes may be written as follows. If $\eta_z \geq \eta_y$ then, for the nominal curvature method,

$$M_{Edz} = M_{0ez} + Ne_{iy} + M_{2z}$$

$$M_{Edy} = M_{0ey}$$

Or, for the moment multiplier method,

$$M_{Edz} = (M_{0ez} + Ne_{iy})[1 + \beta/(N_B/N_{ED} - 1)]$$

$$M_{Edy} = M_{0ey}$$

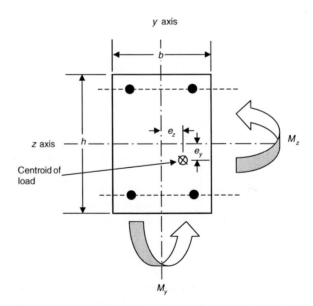

Fig. 7.10. Notation for biaxial bending

If $\eta_y > \eta_z$ then, for the nominal curvature method,

$$M_{Edz} = M_{0ez}$$

$$M_{Edy} = M_{0ey} + Ne_{iz} + M_{2y}$$

Or, for the moment multiplier method,

$$M_{Edz} = M_{0ez}$$

$$M_{Edy} = (M_{0ey} + Ne_{iz})[1 + \beta/(N_B/N_{ED} - 1)]$$

Note that, if the first-order bending moment only acts about one axis, then either M_{0ez} or M_{0ey} may be zero.

The conditions set out above may lead to a necessity to design the column section for biaxial bending. Where this is necessary, the provisions of *clause 5.8.9.4* may be used. These are discussed in Section 5.2.6.

Clause 5.8.9.4

EN 1992-1-1 does, however, define situations where a simplified approach whereby checks on the basis of uniaxial bending may be carried out for the two axes independently. These are cases where the first-order bending is close to being uniaxial at the critical section for buckling. This condition is defined in *clauses 5.8.9(2)* and *5.8.9(3)* as being when

Clause 5.8.9(2)
Clause 5.8.9(3)

(1) $\lambda_y/\lambda_z \leq 2$ and $\lambda_z/\lambda_y \leq 2$
(2) either $(e_y/h)/(e_z/b) \leq 0.2$ or $(e_z/h)/(e_y/b) \leq 0.2$.

7.3.6. Global second-order effects

The body of EN 1992-1-1 only deals in detail with the design of isolated columns for second-order effects. This is the commonest situation where an individual column may deflect to failure while the remainder of the structure does not deflect significantly. It is thus assumed that there is no significant sidesway at the time that the column fails. Deflections of the whole structure will, however, occur which will impose sidesway on all vertical members and, if this sidesway is significant, it will need to be considered in the design. *Clause 5.8* does give rules for deciding whether the possible sidesway may be ignored, but gives no methods for dealing with the problem if sidesway is significant. Methods are, however, given in informative *Annex D*. Being an informative annex, *Annex D* may be adopted as normative by individual countries, or they may introduce their own methods within the National Annex.

Clause 5.8

The rules by which it can be decided that global second-order effects may be ignored are given in *clauses 5.8.2(6)* and *5.8.3.3*. *Clause 5.8.2(6)* states the general principle that second-order effects may be ignored if they are less than 10% of the corresponding first-order effects. This is a useful principle to have stated, but is practically unhelpful. *Clause 5.8.3.3* provides an operational rule which states that global second-order effects may be ignored if

Clause 5.8.2(6)
Clause 5.8.3.3

$$F_{V, Ed} \leq k_1 n_s/(n_s + 1.6) \times \sum E_{cd} I_c/L^2$$

where $F_{V, Ed}$ is the total vertical load on both braced and bracing elements, n_s is the number of storeys, L is the total height of the structure above the level of moment restraint, I_c is the second moment of area calculated on the basis of the concrete section of the members considered to support lateral loads, and E_c is the design modulus of elasticity of the concrete.

In general, it is likely that it can be established by inspection whether or not global second-order effects need consideration.

The method given in *Annex D* for designing for global second-order effects will not be considered in detail here as the description in the annex seems clear. Basically, the method calculates a magnified horizontal load for which the structure should be designed.

7.3.7. Walls

Reinforced concrete walls may be treated by the either the moment multiplier method or the nominal curvature method in exactly the same way as columns except that it is only necessary to consider the possibility of deformation about the minor axis (i.e. only deformation perpendicular to the plane of the wall need be considered). The approach for the design of walls assumes that the first-order analysis for the vertical and lateral loads will give a distribution of axial load along the wall. Consideration of the interaction of the floors and beams framing into the walls will lead to transverse first-order moments. Sections of wall will then be designed for the maximum axial load per unit length of wall within the section considered combined with any transverse moments and the appropriate accidental and second-order eccentricities.

7.3.8. Lateral buckling of slender beams

Clause 5.9

Where a beam is narrow compared with its span or depth, it may fail by lateral buckling of the form illustrated schematically in Fig. 7.11. Beams where this is likely to be a problem are relatively rare, and hence a simple, conservative check is normally sufficient. EN 1992-1-1 includes such a check in *clause 5.9*. This states that the safety will be adequate without further check provided that:

(1) For persistent situations,

$$l_{0t}/b \le 50/(h/b)^{1/3} \qquad h/b \le 2.5$$

(2) For transient situations,

$$l_{0t}/b \le 70/(h/b)^{1/3} \qquad h/b \le 3.5$$

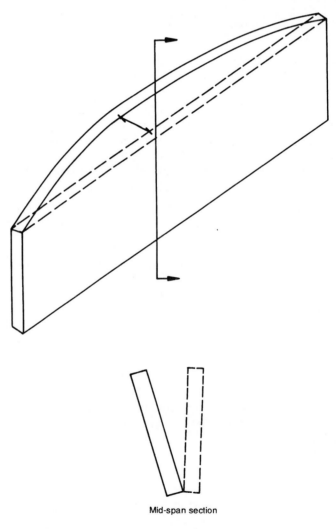

Mid-span section

Fig. 7.11. Buckling of a slender beam

Fig. 7.12. Cross-section of a slender beam with a parapet

where l_{0t} is the distance between torsional restraints, h is the overall depth of the beam in the central part of l_{0t} and b is the width of the compression zone or flange.

No indication is given in EN 1992-1-1 as to how a further check should be formulated, should this be necessary. The CEB–FIP 1990 Model Code does, however, give a more detailed method of analysis in its clause 6.7.3.3.4. This method has similarities with the nominal curvature method in that it postulates an ultimate deflected shape and then ensures that the critical section can withstand the resulting internal actions. Reference should be made to the CEB–FIP 1990 Model Code for details of the method.

A problem which occasionally occurs in practice with slender beams is where, for example, an edge beam is designed with a thin parapet cast monolithically as sketched in Fig. 7.12. Such a beam would normally be designed ignoring the effect of the upstand parapet; however, rigorous interpretation of rules such as those in EN 1992-1-1 would imply that such a member cannot be used because of the slenderness of the parapet. It must, in such circumstances be satisfactory to state that if the beam is adequately safe without the parapet, then the addition of the parapet cannot make it less safe.

CHAPTER 8

Serviceability limit states

8.1. General

EN 1992-1-1 deals in some detail with three common serviceability limit states. These are:

- Limitation of stresses
- Control of cracking
- Control of deflections

Clause 7.2
Clause 7.3
Clause 7.4

Design for any limit state requires the definition of four quantities. These are:

(1) definition of the appropriate loading and methods of analysis so that the design load effects can be established
(2) definition of the design material properties to be assumed in the verification
(3) definition of criteria defining the limit of satisfactory performance
(4) definition of suitable methods by which performance may be predicted.

Of these, only (3) and (4) will be found in *Section 7* of EN 1992-1-1; (1) will be found in EN 1990 (Section 6 and clause 2.5.3.1) and (2) in *Section 3*. All these matters will be summarized in this chapter. It should be noted that in most cases it will not be necessary to carry out explicit calculations for the serviceability limit states, as simple, 'deemed to satisfy' procedures are given in the code for dealing with all three of the limit states covered. This approach is acceptable because serviceability is intrinsically less critical than the ultimate limit states, and major calculation effort is not justified. For example, if the structure is wrongly designed and the strength is even 1% below the imposed loads, the structure collapses. By comparison, if the crack width turns out to be 0.33 mm instead of 0.3 mm, nothing more serious than some grumbling from the owner and a cosmetic repair is likely to result.

8.1.1. Assessment of design action effects

EN 1990 (clause 6.5.3) defines three combinations of actions which may need to be considered when designing for a serviceability limit state. These are:

- Characteristic combination:

$$G_{k,j} + (P) + Q_{k,1} + \sum \psi_{0,i} Q_{k,i} \tag{D8.1}$$

- Frequent combination:

$$G_{k,j} + (P) + Q_{1,1} + \sum \psi_{2,i} Q_{k,i} \tag{D8.2}$$

- Quasi-permanent combination:

$$G_{k,j} + (P) + \sum \psi_{2,i} Q_{k,i} \tag{D8.3}$$

It will be understood from these formulae that the partial factor on the loads is always 1.0 for serviceability limit states. For buildings, the following simplifications which may be used for the characteristic or frequent combinations:

- Where there is only one variable action:

$$G_{k,j} + (P) + Q_k$$

- Where there are two or more variable actions:

$$G_{k,j} + (P) + 0.9\sum Q_{k,i}$$

Analysis may be elastic (without redistribution). This analysis may normally be based on the stiffness of the uncracked section; however, if it is suspected that cracking may have a significant unfavourable effect on the performance, then a more realistic analysis taking account of the cracking should be used. This possibility may be ignored for normal building structures.

Clause 5.10.9   The determination of the effects of prestress is covered in *clause 5.10.9*.

8.1.2. Material properties
Clause 6.5.4 of EN 1990 states that the partial safety factor applied to material properties should generally be 1.0.

The properties of the materials which are normally significant in serviceability calculations are the modulus of elasticity of the reinforcement, the modulus of elasticity, the creep coefficient, the shrinkage strain and the tensile strength of the concrete. All this information is given in *Section 3*, but is summarized here for convenience.

Modulus of elasticity of reinforcement

Clause 3.2.7(4)   For ordinary reinforcement this may be taken as 200 kN/mm² (*clause 3.2.7(4)*). For prestressing
Clause 3.3.6(2)   steel, 205 kN/mm² may be assumed for wires and bars, and 195 kN/mm² for strand (*clauses
Clause 3.3.6(3)   3.3.6(2)* and *3.3.6(3)*).

Modulus of elasticity of concrete

Clause 3.1.3   Values of the modulus of elasticity are given in *clause 3.1.3* and *Table 3.1*. The elastic modulus of concrete varies with other factors than just the strength (e.g. the aggregate type), and, if an accurate prediction of serviceability conditions is required, it will be necessary to establish the value of E_{cm} by tests on the type of concrete actually being used.

Creep coefficient

The creep coefficient, which is defined as the ratio of the creep deformation to the instantaneous elastic deformation, depends upon many factors, the most significant of which are the age at loading, the time under load, the relative humidity, the section geometry, the concrete strength and the type of cement. Other factors such as the aggregate type may also be significant. As with the modulus of elasticity, the only way of determining the creep performance of concrete with any real reliability is to obtain data from creep tests on the concrete actually being used. This information is rarely available, and more approximate means have to be adopted. The equations given in EN 1992-1-1 should be viewed in this light. Full equations for the prediction of the creep coefficient are given in *Annex B*. These have
Clause 3.1.4   been used to produce design graphs (*Fig. 3.1* in *clause 3.1.4*).

Free shrinkage strain

The free shrinkage strain depends upon the same basic variables as creep, and equations are given in *Annex B* for its prediction. Simplified values are given in *clause 3.1.4*.

Tensile strength of concrete

Table 3.1 gives the mean tensile strength and the upper and lower characteristic tensile strengths as a function of the characteristic compressive strength of the concrete. Equations are also given in *Table 3.1*.

8.2. Limitation of stresses under serviceability conditions

8.2.1. General

The question of the possible limitation of stresses under serviceability loads was the subject of a great deal of discussion during the drafting of EN 1992-1-1. The basic reasons for differences of opinion arose from the different way in which design methods had developed in different countries and a natural desire in each country that the new and untried Eurocode should not force design too far from existing experience in that country. Limitations on the compressive stress in the concrete were particularly contentious. In many countries, section design for both prestressed and reinforced concrete has been carried out on the basis of ultimate strength methods, and no explicit check has been required on the concrete stresses under service loads. These countries were determined that no such checks should be introduced since these could be shown to have a significant effect on the economy of some types of member, notably columns. Other countries have not so far adopted ultimate load methods, and still base their design on an elastic analysis of sections combined with a stress limit. While these countries were prepared to accept ultimate strength methods for section design, they also wanted to retain elastic design as an additional requirement to ensure that member sizes, particularly columns, did not become much smaller or more lightly reinforced than those arising from their current design methods. The check on the compressive stress in the concrete at the serviceability limit state provided them with this safeguard.

In order to develop a rational draft for this section of the code which would have some hope of general acceptance, an attempt was made to establish what rational basis existed for a limit on stresses under serviceability conditions. Two limits need to be considered: compressive stress limits and tensile stress limits. These will be considered in turn.

Limits to compressive stress

Two reasons are usually put forward for limiting compressive stresses in the concrete. These are (1) to avoid the formation of microcracks in the concrete which might reduce durability and (2) to avoid excessive creep. There is some logic in both reasons. It is commonly accepted that microcracking will start to develop in concrete when the compressive stress exceeds about 70% of the compressive strength. However, none of the established publications on durability suggest any relationship between stress level, microcracking and durability problems (e.g. see *CEB Bulletin d'Information 185*[9]). EN 1992-1-1 suggests that, in aggressive environments (exposure classes XD, XF and XS), it may be reasonable to limit the stress to $0.6f_{ck}$ if other measures, such as an increase to the cover to any compressive reinforcement or the provision of confinement to the compressed concrete by transverse reinforcement.

The design methods used for checking serviceability assume that creep will be proportional to stress. This assumption is increasingly invalid as the stress increases above about 50% of the compressive strength. It could therefore be logical to impose a limit on compressive stress under the quasi-permanent combination of loads where a higher level of creep than might be predicted could seriously impair performance. This problem has been dealt with in *clause 3.1.4*, where a correction equation has been introduced which enables the creep to be calculated for levels of compressive stress above $0.45f_{ck}$.

Clause 3.1.4

It will be seen that there is no case for a blanket limitation to compressive stresses but that, in particular cases, it may be prudent to apply such limits if the effects of non-linear creep are not explicitly calculated. EN 1992-1-1 (*clause 7.2*) suggests that non-linear creep should be taken into account if the compressive stress exceeds $0.45f_{ck}$ under the quasi-permanent load.

Clause 7.2

In UK practice, compressive stresses have been checked in the design of prestressed members where high compressive stresses are likely to be present permanently, but no check has been required in the design of reinforced sections over the last 50 years. There has been no evidence of problems with performance that could be related to high compressive stress levels in service. It therefore seems reasonable to assume that there is no general requirement to check the compressive stresses in reinforced-concrete members under service conditions, although there may be special circumstances where this might be prudent.

Limitation of tensile stresses

Limits may be defined for the tensile stresses in the concrete or in the reinforcement. In the case of reinforcement, it seems reasonable to ensure that inelastic deformations of the steel are avoided under service loads. Such deformations would, firstly, invalidate any calculations of cracking or deflections which assume that the reinforcement behaves elastically and, secondly, could result in excessively large cracks. This could be particularly critical in areas subjected to frequent variations in loading. The limits suggested in *clause 7.2(5)* are aimed at ensuring that inelastic deformations do not occur, and include some margin of safety against this occurring.

Clause 7.2(5)

Prestressing steels are much more sensitive to corrosion damage than are normal reinforcing bars. This is partly due to differences in metallurgy but mostly due to their smaller diameter and the high level of stress at which they normally operate. A prestressing tendon may frequently operate for its whole life at a stress of up to 70% of its tensile strength over its whole length. A reinforcing bar is unlikely to be stressed above about 50% of its yield strength for any significant period, and then only in a very limited area close to sections of maximum moment. It is therefore necessary to take much more stringent precautions against corrosion with prestressed members than with reinforced members. In any parts of a member which could be exposed to salt, it is necessary to ensure that no direct paths exist which could give direct access for the salt to the surface of the tendon. Cracks could provide such a path, and so cracks which could penetrate to the tendons should be avoided in salty environments. This objective can conveniently be achieved by ensuring that no tension develops in the concrete surrounding the tendons. *Clause 7.2(5)* provides an element of extra safety by requiring that, in environments where this precaution is necessary, the tendon is surrounded by at least 25 mm of uncracked concrete.

Clause 7.2(5)

8.2.2. Procedure for stress checks

In general, stress checks may be avoided for reinforced concrete building structures if the advice given in the National Annex is deemed relevant.

If stresses are to be checked, the calculation is done on the basis of the following assumptions:

(1) Plane sections remain plane.
(2) Reinforcement and concrete in tension are assumed to be elastic.
(3) Concrete is assumed to be elastic up to its tensile strength, f_{ctm}. If the stress has exceeded f_{ctm}, the section is assumed to be cracked, and the concrete in tension is ignored.
(4) EN 1992-1-1 does not give details of how creep under varying loads should be handled. However, a reasonable approach seems to be to assume that creep may generally be taken into account by assuming that the ratio of the elastic modulus for steel to that for concrete (modular ratio) is 15. A lower value, based on the actual elastic modulus of the concrete may be used where less than 50% of the stresses arise from quasi-permanent loads. A more accurate assessment of the effects of creep may, of course, be used if desired.

The code states that shrinkage and temperature stresses should be taken into account where these are likely to be significant. This will not usually be necessary for normal reinforced and prestressed members in buildings; however, appropriate methods for doing this will be considered on p. 148.

In general, stress checks can be carried out using standard elastic formulae. Thus, for beams subjected only to flexure, stresses may be calculated from

$$\text{stress} = My/I \tag{D8.4}$$

where M is the applied moment, y is the distance from the neutral axis to the point considered and I is the second moment of area of the section (this should be based on either a cracked or uncracked section, as appropriate).

Table 8.1. Neutral axis depths and moments of inertia for flanged beams

(a) $hf/d = 0.2$

	b_r/b									
	1.0		0.5		0.4		0.3		0.2	
α_{ep}	x/d	I/bd^3	x/d	I/bd^3	x/d	I/bd^3	x/d	I/bd^3	x/d	I/bd^3
0.020	0.181	0.015								
0.030	0.217	0.022	0.217	0.022	0.217	0.022	0.217	0.022	0.217	0.022
0.040	0.246	0.028	0.248	0.028	0.248	0.028	0.249	0.028	0.249	0.028
0.050	0.270	0.033	0.274	0.033	0.275	0.033	0.276	0.033	0.278	0.033
0.060	0.292	0.038	0.298	0.038	0.300	0.038	0.302	0.038	0.304	0.038
0.070	0.311	0.043	0.320	0.043	0.322	0.043	0.325	0.043	0.327	0.043
0.080	0.328	0.048	0.340	0.047	0.343	0.047	0.346	0.047	0.349	0.047
0.090	0.344	0.052	0.358	0.052	0.361	0.052	0.365	0.051	0.369	0.051
0.100	0.358	0.057	0.375	0.056	0.379	0.056	0.383	0.055	0.388	0.055
0.110	0.372	0.061	0.390	0.060	0.395	0.059	0.400	0.059	0.406	0.059
0.120	0.384	0.064	0.405	0.063	0.410	0.063	0.416	0.063	0.422	0.062
0.130	0.396	0.068	0.418	0.067	0.424	0.066	0.430	0.066	0.437	0.065
0.140	0.407	0.072	0.431	0.070	0.437	0.070	0.444	0.069	0.452	0.069
0.150	0.418	0.075	0.443	0.073	0.450	0.073	0.457	0.072	0.466	0.071
0.160	0.428	0.078	0.455	0.076	0.462	0.076	0.470	0.075	0.478	0.074
0.170	0.437	0.082	0.466	0.079	0.473	0.078	0.481	0.078	0.491	0.077
0.180	0.446	0.085	0.476	0.082	0.484	0.081	0.493	0.080	0.502	0.079
0.190	0.455	0.088	0.486	0.085	0.494	0.084	0.503	0.083	0.513	0.082
0.200	0.463	0.091	0.495	0.087	0.504	0.086	0.513	0.085	0.524	0.084
0.210	0.471	0.094	0.504	0.090	0.513	0.089	0.523	0.088	0.534	0.086
0.220	0.479	0.096	0.513	0.092	0.522	0.091	0.532	0.090	0.543	0.089
0.230	0.486	0.099	0.521	0.094	0.531	0.093	0.541	0.092	0.552	0.091
0.240	0.493	0.102	0.529	0.097	0.539	0.095	0.549	0.094	0.561	0.093

(b) $hf/d = 0.3$

	b_r/b									
	1		0.5		0.4		0.3		0.2	
p	x/d	I/bd^3	x/d	I/bd^3	x/d	I/bd^3	x/d	I/bd^3	x/d	I/bd^3
0.020	0.181	0.015								
0.030	0.217	0.022								
0.040	0.246	0.028								
0.050	0.270	0.033								
0.060	0.292	0.038								
0.070	0.311	0.043	0.311	0.043	0.311	0.043	0.311	0.043	0.311	0.043
0.080	0.328	0.048	0.328	0.048	0.329	0.048	0.329	0.048	0.329	0.048
0.090	0.344	0.052	0.345	0.052	0.345	0.052	0.345	0.052	0.346	0.052
0.100	0.358	0.057	0.360	0.056	0.361	0.056	0.361	0.056	0.362	0.056
0.110	0.372	0.061	0.375	0.060	0.375	0.060	0.376	0.060	0.377	0.060
0.120	0.384	0.064	0.388	0.064	0.389	0.064	0.390	0.064	0.391	0.064
0.130	0.396	0.068	0.401	0.068	0.402	0.068	0.403	0.068	0.404	0.068
0.140	0.407	0.072	0.413	0.071	0.414	0.071	0.416	0.071	0.417	0.071
0.150	0.418	0.075	0.425	0.075	0.426	0.075	0.428	0.075	0.430	0.075
0.160	0.428	0.078	0.436	0.078	0.437	0.078	0.439	0.078	0.441	0.078
0.170	0.437	0.082	0.446	0.081	0.448	0.081	0.450	0.081	0.452	0.081
0.180	0.446	0.085	0.456	0.084	0.458	0.084	0.461	0.084	0.463	0.084

Table 8.1. (Contd)

	b_r/b									
	1.0		0.5		0.4		0.3		0.2	
p	x/d	I/bd^3	x/d	I/bd^3	x/d	I/bd^3	x/d	I/bd^3	x/d	I/bd^3
0.190	0.455	0.088	0.466	0.087	0.468	0.087	0.471	0.087	0.473	0.087
0.200	0.463	0.091	0.475	0.090	0.477	0.090	0.480	0.090	0.483	0.089
0.210	0.471	0.094	0.483	0.093	0.486	0.092	0.489	0.092	0.493	0.092
0.220	0.479	0.096	0.492	0.095	0.495	0.095	0.498	0.095	0.502	0.095
0.230	0.486	0.099	0.500	0.098	0.503	0.098	0.507	0.097	0.511	0.097
0.240	0.493	0.102	0.508	0.100	0.511	0.100	0.515	0.100	0.519	0.099

(c) $hf/d = 0.4$

	b_r/b									
	1.0		0.5		0.4		0.3		0.2	
p	x/d	I/bd^3	x/d	I/bd^3	x/d	I/bd^3	x/d	I/bd^3	x/d	I/bd^3
0.020	0.181	0.015								
0.030	0.217	0.022								
0.040	0.246	0.028								
0.050	0.270	0.033								
0.060	0.292	0.038								
0.070	0.311	0.043								
0.080	0.328	0.048								
0.090	0.344	0.052								
0.100	0.358	0.057								
0.110	0.372	0.061								
0.120	0.384	0.064								
0.130	0.396	0.068								
0.140	0.407	0.072	0.407	0.072	0.407	0.072	0.407	0.072	0.407	0.072
0.150	0.418	0.075	0.418	0.075	0.418	0.075	0.418	0.075	0.418	0.075
0.160	0.428	0.078	0.428	0.078	0.428	0.078	0.428	0.078	0.428	0.078
0.170	0.437	0.082	0.438	0.082	0.438	0.082	0.438	0.082	0.438	0.082
0.180	0.446	0.085	0.447	0.085	0.447	0.085	0.448	0.085	0.448	0.085
0.190	0.455	0.088	0.456	0.088	0.457	0.088	0.457	0.088	0.457	0.088
0.200	0.463	0.091	0.465	0.091	0.465	0.091	0.466	0.091	0.466	0.091
0.210	0.471	0.094	0.473	0.094	0.474	0.094	0.474	0.094	0.474	0.093
0.220	0.479	0.096	0.481	0.096	0.482	0.096	0.482	0.096	0.483	0.096
0.230	0.486	0.099	0.489	0.099	0.490	0.099	0.490	0.099	0.491	0.099
0.240	0.493	0.102	0.496	0.101	0.497	0.101	0.498	0.101	0.498	0.101

Similarly, for uncracked reinforced members subject to moments and axial load or prestressed beams, the stresses may be calculated from the standard relationship

$$\text{stress} = My/I + N/A_c \tag{D8.5}$$

where N is the applied axial force and A_c is the area of the total concrete section.

Tables 8.1 and 8.2 give values for the neutral axis depth and second moments of areas of cracked and uncracked rectangular and flanged sections. For cracked sections, only singly reinforced members are considered. Figures 8.1 and 8.2 give values for the cracked neutral axis depths and second moments of area of doubly reinforced rectangular sections; these have been taken from Rowe.[10]

Table 8.2. Properties of uncracked sections

| | h_f/h | | | | | | | |
| | 1.0 | | 0.4 | | 0.3 | | 0.2 | |
b_r/b	x/h	I/bh^3	x/h	I/bh^3	x/h	I/bh^3	x/h	I/bh^3
0.100			0.265	0.020	0.245	0.019	0.243	0.019
0.150			0.292	0.026	0.280	0.026	0.288	0.026
0.200			0.315	0.032	0.309	0.032	0.322	0.031
0.250			0.336	0.037	0.334	0.037	0.350	0.036
0.300	0.500	0.083	0.355	0.042	0.356	0.042	0.373	0.041
0.350			0.372	0.046	0.375	0.046	0.392	0.045
0.400			0.388	0.050	0.391	0.050	0.408	0.049
0.450			0.401	0.054	0.406	0.054	0.421	0.052
0.500			0.414	0.057	0.419	0.057	0.433	0.055

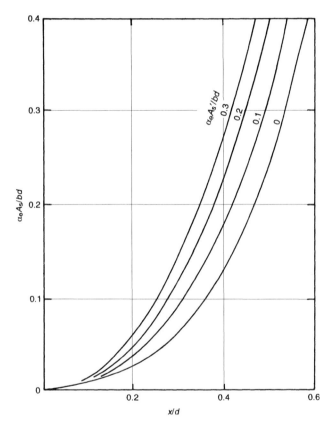

Fig. 8.1. Neutral axis depths calculated on the basis of a cracked section

The calculation of stresses in a cracked member subject to axial forces and moments is less simple since the second moment of area of the cracked section is a function of the axial force. In such cases it is generally easier to write out the equations of equilibrium and solve these iteratively. An approximate calculation of the steel stress in such sections can, however, be made by the use of the design chart in Fig. 8.3. This is used as follows:

(1) Calculate the average axial stress in the section by dividing the longitudinal force by the area of the concrete section. This is then divided by the reinforcement ratio $(A_s/b_w d)$ and multiplied by the factor b_w/b, to give the longitudinal force factor, $Nb_w/A_c b$.

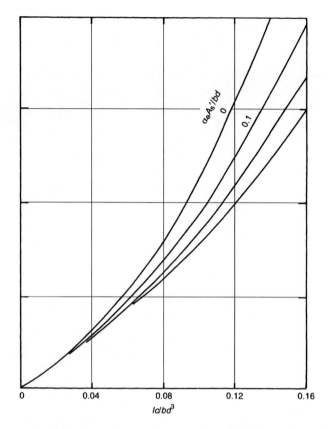

Fig. 8.2. Second moments of the area of rectangular sections calculated on the basis of a cracked section

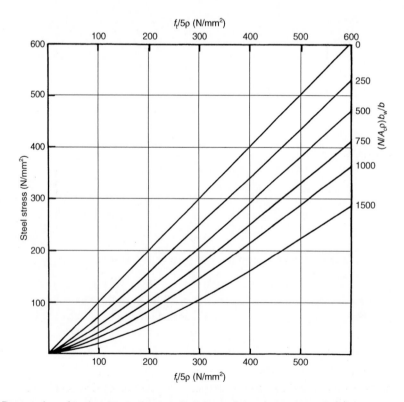

Fig. 8.3. Design chart for the approximate calculation of the stress in reinforcement in a prestressed beam

(2) Calculate the 'notional tensile stress', f_t, at the tension face of the section. This is calculated on the basis of an uncracked section, even though the section is cracked. This is then used to calculate the factor $f_t/5$.

(3) The above two factors can then be used in Fig. 8.3 to give the steel stress in any unprestressed reinforcement in a prestressed beam or the change in stress in the prestressing tendons from the condition where the stress in the concrete immediately surrounding the steel is zero.

Example 8.1

Calculate the stresses in a T beam with the dimensions given below when subjected to a moment of 100 kNm:

- Overall breadth: 600 mm
- Overall depth: 500 mm
- Effective depth: 450 mm
- Flange depth: 150 mm
- Rib breadth: 200 mm
- Steel area: 2455 mm^2

The concrete grade is C30/37.

From *Table 3.1*, the mean tensile strength of the concrete is 2.8 N/mm^2. From Table 8.2 the neutral axis depth and second moment of area of the uncracked section can be found to be

$$x_{\mathrm{I}} = 0.367 \times 500 = 184 \text{ mm}$$

$$I_{\mathrm{I}} = 0.045 \times 600 \times 500^3 = 3375 \times 10^6 \text{ mm}^4$$

The tensile stress at the bottom of the section assuming an uncracked section is thus given by

$$f_t = 100 \times (500 - 184)/3375 = 9.36 \text{ N/mm}^2$$

This is greater than the tensile strength of the concrete, so the section should be assumed to be cracked. The neutral axis depth and second moment of area of the cracked section can be obtained by interpolation from Tables 8.1b and 8.1c. The reinforcement ratio multiplied by the modular ratio is given by

$$\rho = 2455/600/450 \times 15 = 0.136$$

From the tables,

$$x_{\mathrm{II}} = 0.41 \times 450 = 185 \text{ mm}$$

$$I_{\mathrm{II}} = 0.07 \times 600 \times 450^3 = 3827 \times 10^6 \text{ mm}^4$$

The stresses can now be calculated:

compressive stress in the concrete $= 100 \times 185/3827 = 4.83$ N/mm^2

tensile stress in steel $= 100 \times (450 - 185) \times 15/3827 = 104$ N/mm^2

Both these stresses are well below the limits given in the code.

Stress checks for reinforced concrete columns

The basic approach for checking the stresses in columns is the same as for beams and, for uncracked members, the formulae given in Section 8.2.2 above for combined axial force and bending may be used. For cracked sections, the algebra becomes rather more involved, and the design charts given in Figs 8.4a–x may be used. The first set of charts (Figs 8.4a–l) give the reinforcement areas necessary to meet the limit of $0.6f_{ck}$ while the second set (Figs 8.4m–x) give the reinforcement areas necessary to meet the limit of $0.45f_{ck}$.

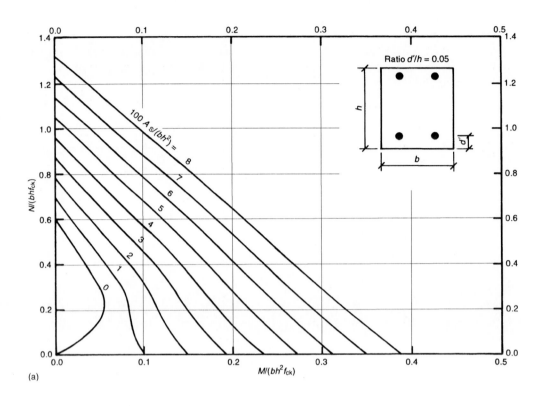

(a)

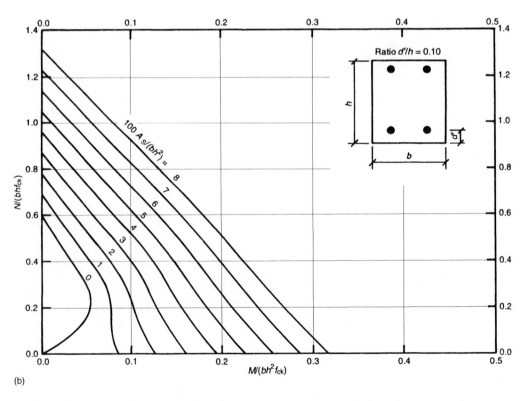

(b)

Fig. 8.4. Design charts for ensuring that the compressive stress in reinforced concrete columns does not exceed $0.6f_{ck}$

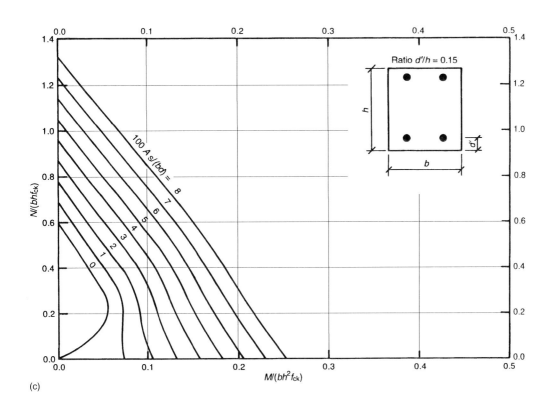

(c)

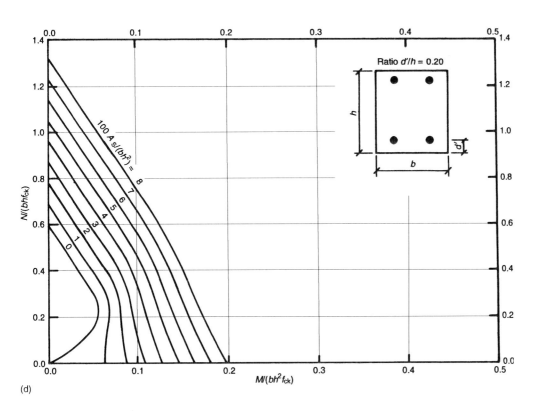

(d)

Fig. 8.4. (Contd)

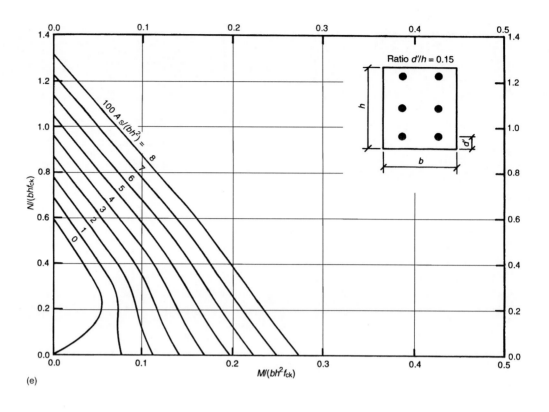

(e)

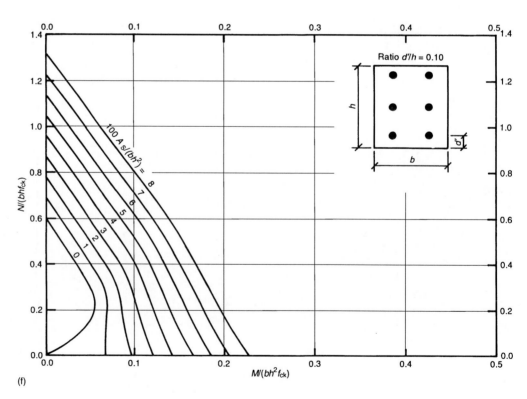

(f)

Fig. 8.4. (Contd)

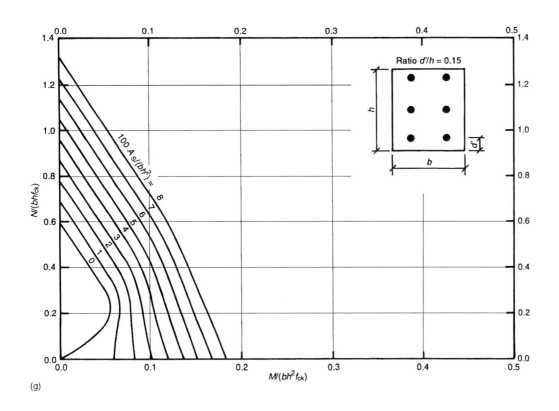

(g)

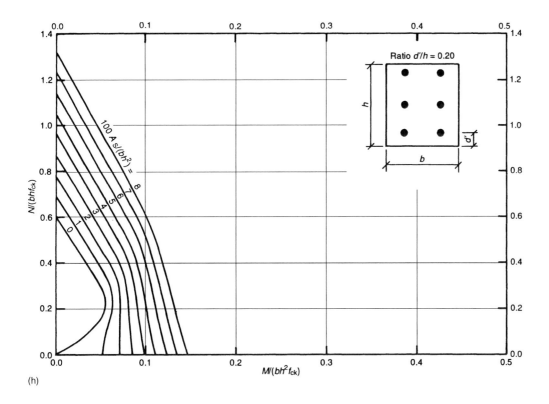

(h)

Fig. 8.4. (Contd)

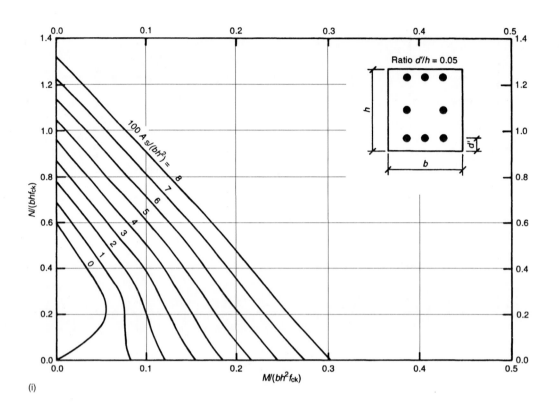

(i)

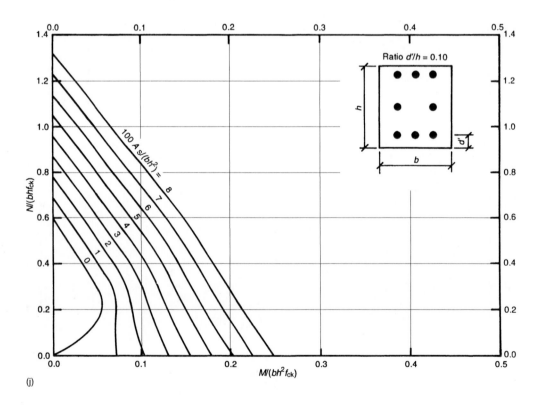

(j)

Fig. 8.4. (Contd)

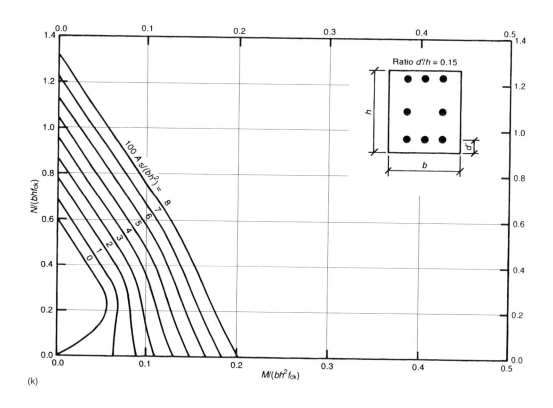

(k)

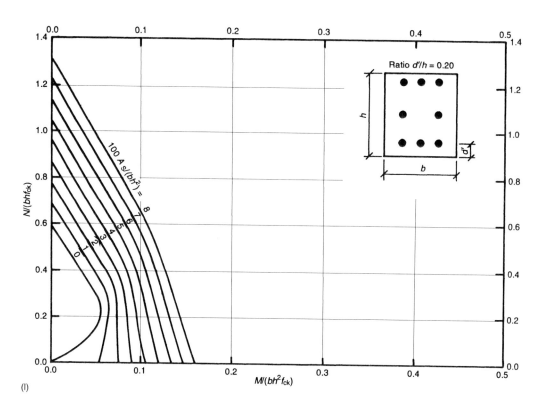

(l)

Fig. 8.4. (Contd)

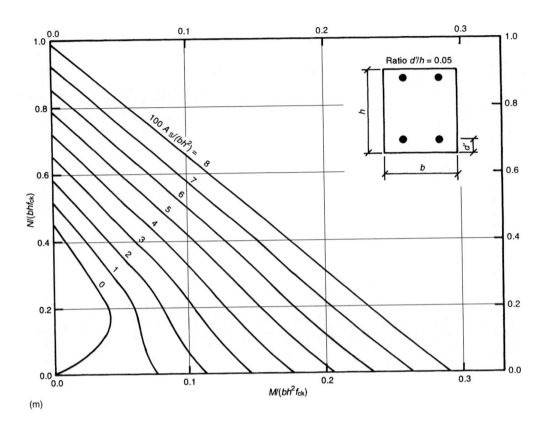

(m)

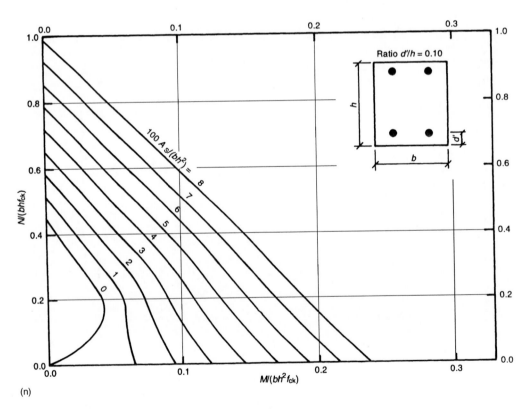

(n)

Fig. 8.4. (Contd)

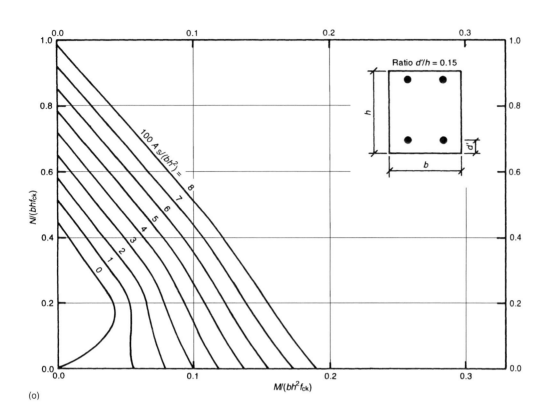

(o)

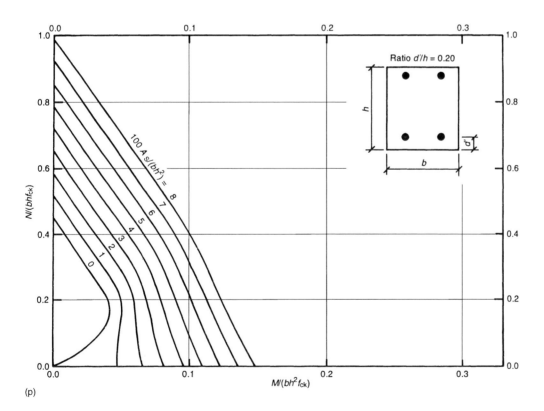

(p)

Fig. 8.4. (Contd)

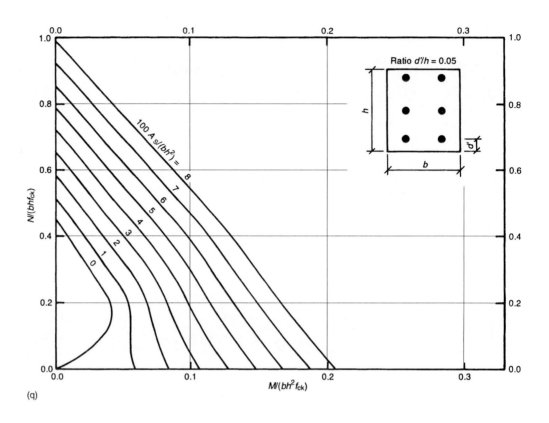

(q)

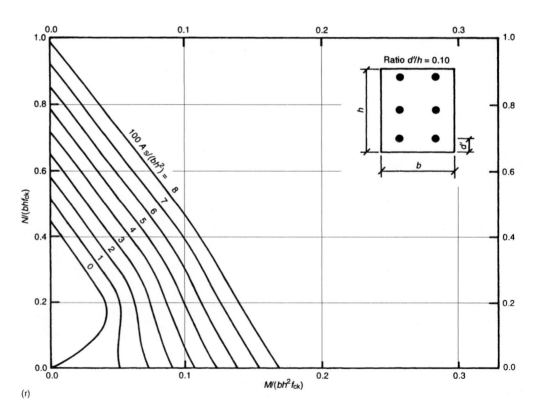

(r)

Fig. 8.4. (Contd)

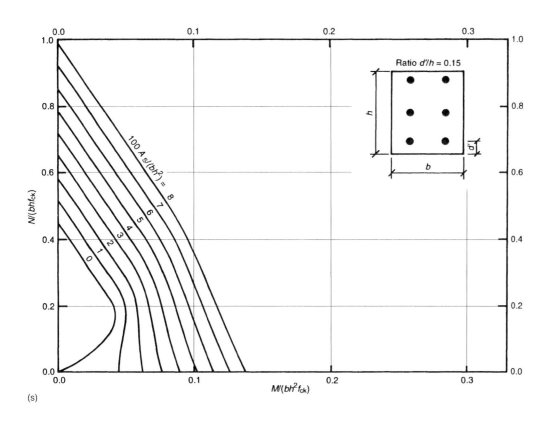

(s)

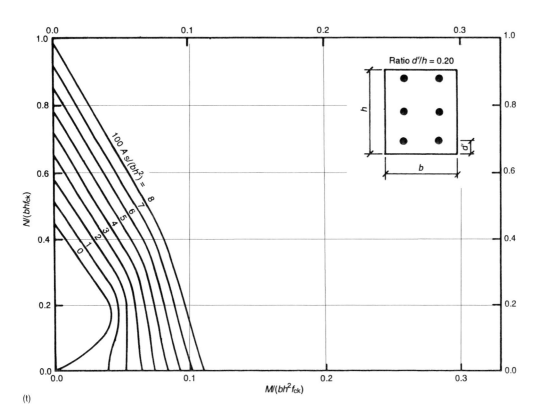

(t)

Fig. 8.4. (Contd)

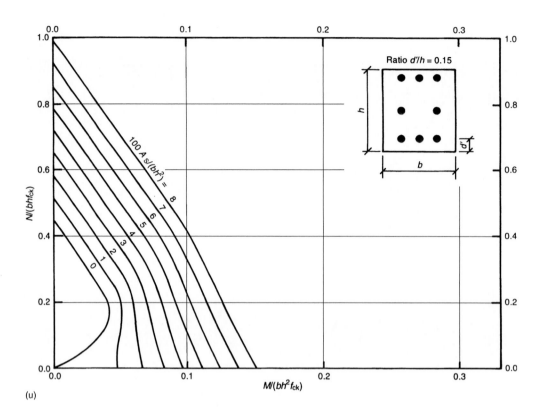

(u)

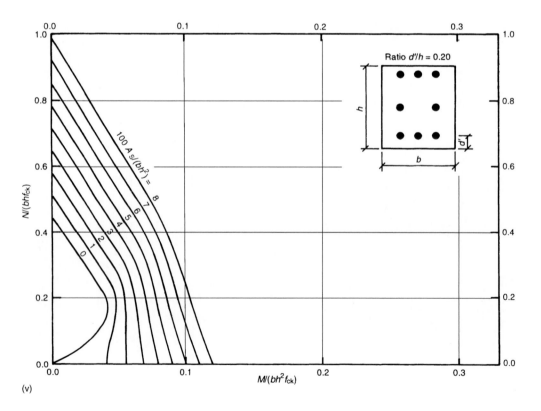

(v)

Fig. 8.4. (Contd)

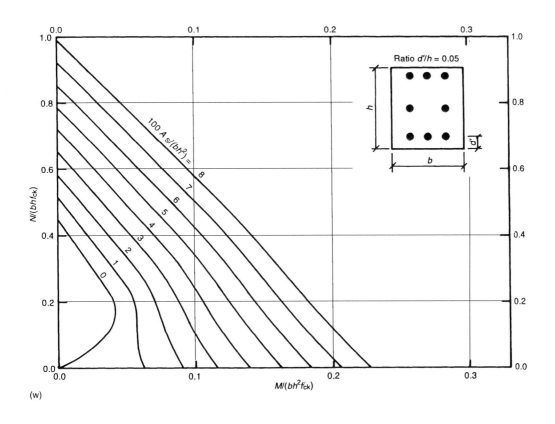

(w)

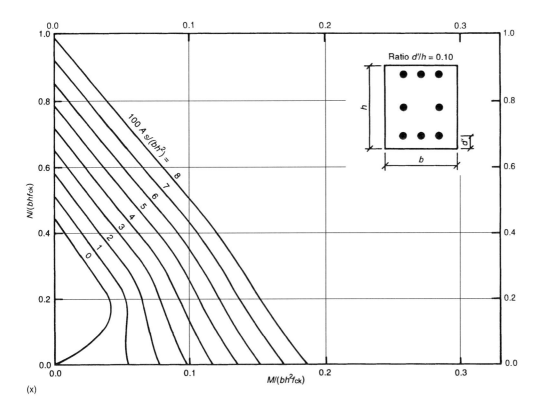

(x)

Fig. 8.4. (Contd)

Stresses due to temperature changes

A problem which occasionally has to be dealt with in deeper members is the estimation of the stresses induced in a member due to non-uniform temperature distributions. For members which are uncracked and unrestrained, well-established methods of analysis exist. Under the influence of non-uniform and non-linear temperature changes, the member will change in overall length and will also bend. The overall length change will be proportional to the average temperature change. For any general section subject to any general distribution of temperature, the average shortening will be given by

$$\varepsilon_{av} = \alpha/A\int_0^h b_y t_y \, dy \qquad \qquad \text{(D8.6)}$$

where A is the total area of the section, h is the overall depth of the section, b_y is the breadth of the section at depth y, t_y is the temperature change at depth y and α is the coefficient of thermal expansion.

The curvature can be calculated by taking moments about the section centroid. This gives

$$1/r = \alpha/I\iint b_y t_y (y - Y)\,dy \qquad \qquad \text{(D8.7)}$$

where Y is the depth to the centroid of the section and I is the second moment of area of the section.

The actual strain at level y is now given by

$$\varepsilon_{ay} = \varepsilon_{av} + 1/r(y - Y) \qquad \qquad \text{(D8.8)}$$

The stress at any level can now be obtained from the difference between the actual strain and the free temperature movement at that level. Thus,

$$\sigma_y = E_c(\alpha t_y - \varepsilon_{ay}) \qquad \qquad \text{(D8.9)}$$

While these equations may look very complicated, they can, in fact, be solved quite easily by numerical methods. A very simple approach is to use a spreadsheet program.

The above method has been developed assuming that the member considered is unrestrained. The effects of restraint can be introduced into the analysis by introducing springs at the member ends which will resist axial and rotational movement. An iterative calculation is then required to develop a situation giving equilibrium between the spring forces and the forces in the member.

A similar approach can be adopted to deal with cracked sections subjected to temperature change. The method becomes highly iterative, and is beyond the scope of this guide. In general, cracking, by reducing the stiffness of the member, will lead to a major relaxation in the stresses induced. It will therefore generally be satisfactory to ignore temperature-induced stresses in cracked members.

8.3. Control of cracking

8.3.1. General

Clause 7.3

This section covers the material in *clause 7.3* of EN 1992-1-1.

Crack width limits

There are many reasons for wishing to limit the widths of cracks to a relatively low value. Among the most commonly cited reasons are:

- to avoid possible corrosion damage to the reinforcement due to deleterious substances penetrating to the reinforcement down the cracks
- to avoid, or limit, leakage through cracks – this is commonly a critical design consideration in water retaining structures
- to avoid an unsightly appearance.

All of the above reasons have been researched to some degree, but no clear definition of permissible crack width has emerged from any of these studies. The following provides a very brief summary of the results from these studies.

- **Cracking and corrosion**. This is the most extensively researched area. Summaries of the findings have been published by a number of authors (e.g. Beeby[11] and Schiessl and Raupach[12]). The development of corrosion is a two-phase process. In fresh concrete the reinforcement is protected from corrosion by the alkaline nature of the concrete. This protection can be destroyed by two mechanisms: carbonation of the concrete to the surface of the reinforcement or ingress of chlorides. Cracks will lead to a local acceleration of both processes by permitting more rapid ingress of either carbon dioxide or chlorides to the surface of the reinforcement. Once the protection provided by the concrete has been destroyed, corrosion can start if the environmental conditions are right. The period from construction up till the initiation of corrosion is usually referred to as the 'initiation phase', while the period after the initiation of corrosion is termed the 'active phase'. The length of the initiation phase is likely to be influenced by crack width. However, this period is likely to be short at a crack, and some corrosion can usually be found on the bar surface where a crack reaches a bar after as little as 2 years even with very small cracks. It is found, however, that this initial corrosion does not develop in cases where the cracks are small or where the bars intersect the cracks. Quite possibly the corrosion products block the cracks and inhibit further corrosion. A more serious situation exists where a crack runs along the line of a bar. There is limited evidence to suggest that, in this case, sustained corrosion may develop in salty environments where the crack width exceeds about 0.3 mm.

  Though less research has been done on the relation between cracking and corrosion in prestressed concrete members, it is generally believed that the risks posed by cracks are greater, and therefore more stringent criteria should be imposed. For this reason, EN 1992-1-1 does not permit cracks to penetrate to the prestressing tendons where the member is exposed to aggressive environments.
- **Leakage**. Only very limited research has been carried out so far into this problem, and this has not led to any agreed basis for crack width limits. Practical experience has suggested that cracks of less than 0.2 mm width which pass right through a section will leak somewhat initially but will quickly seal themselves. This problem is not specifically considered in EN 1992-1-1 as liquid-retaining structures are covered in EN 1992-3.
- **Appearance**. Limited studies suggest that noticeable cracks in structural members cause concern to the occupants of structures, and it is therefore advisable to keep cracks below a width that will not generally be noticed by a casual observer. On a smoothly finished concrete surface, it appears that cracks are unlikely to lead to complaint if the maximum width is kept below 0.4 mm. Clearly, larger widths may be used on rougher forms of surface or where the cracking cannot be seen. This is mentioned in the note to *Table 7.1N*.

The general considerations set out above led to the recommended limits on crack width set out in EN 1992-1-1 in *Table 7.1N*. It should be noted, however, that the information on which the criteria were based is far from being unambiguous. The actual values used in particular countries can be defined in their National Annexes, but the values given in the table have been accepted for buildings in the UK.

Causes of cracking

There are many possible causes of cracking and only a few of these lead to cracks that can be controlled by measures taken during the design. The following are the more common causes.

- Plastic shrinkage or plastic settlement. These are phenomena which occur within the first few hours after casting while the concrete is still in a plastic state. The likelihood of cracks being caused by these phenomena depends upon the bleeding rate of the mix and the evaporation rate. The resulting cracks may be large: up to 2 mm is not uncommon.

- Corrosion. Rust occupies a greater volume than the metal from which it is formed. Its formation therefore causes internal pressures to build up around the bar surface, which will lead to the formation of cracks running along the line of the corroding bars and, eventually, spalling of the concrete cover.
- Expansive chemical reactions within the concrete. Expansive reactions occurring at the concrete surface tend to lead to scaling of the concrete rather than cracking; however, some reactions, such as the alkali–silica reaction, occur within the body of the concrete and can lead to large surface cracks.
- Restrained deformations, such as shrinkage or temperature movements.
- Loading.

Of this list, only the last two causes can be treated by the designer. They are probably the two least serious causes of cracking.

8.3.2. Minimum areas of reinforcement
General principles
An essential condition for the validity of formulae for crack width calculation is that the reinforcement remains elastic. If the reinforcement yields, then the deformation becomes concentrated at the crack where the yielding is occurring, which is not the condition considered in the derivation of the formulae. Particularly critical is the condition immediately after formation of the first crack. Should formation of the first crack lead to yield, then only one single crack will form, and all the deformation will be concentrated at this crack. The principles involved here can be most easily demonstrated by considering a member subjected to pure tension.

The force necessary to cause the member to crack is given by

$$N_r = A_c f_{ct}$$

where N_r is the cracking load, A_c is the area of concrete and f_{ct} is the tensile strength of the concrete.

The strength of the steel is $A_s f_y$. Thus, for the steel not to yield on first cracking, and hence for spread cracking to develop,

$$A_s f_y > A_c f_{ct}$$

or

$$A_s > A_c f_{ct}/f_y$$

This provides the minimum reinforcement area required for controlled cracking. It can be shown that where the cracking is caused exclusively by loading, this limitation is unimportant since no cracks will form under service conditions. However, if tensile stresses may be generated by the restraint of imposed deformations such as shrinkage or thermal movements, then it is essential to ensure that at least the minimum reinforcement area is provided. This result is true for flexure as well as for pure tension. Since it is only in rare cases that it can be said with confidence that there is no possibility of stresses arising from restraint, it is advisable to apply the rules for the minimum steel area in the great majority of practical cases.

While the principles illustrated above for pure tension apply generally, the actual equations will differ for different types of member. To avoid excessive complexity in the rules, a factor k_c is introduced into the above relationship to adjust for different forms of stress distribution.

A further factor, k, is included to allow for the influence of internal self-equilibrating stresses. These arise in cases where deformations of the concrete in the member itself are restrained. The most common forms of such restrained deformation are deformations resulting from either shrinkage or temperature change. These effects do not occur uniformly throughout the section, occurring more rapidly near the member surface. As a result, the

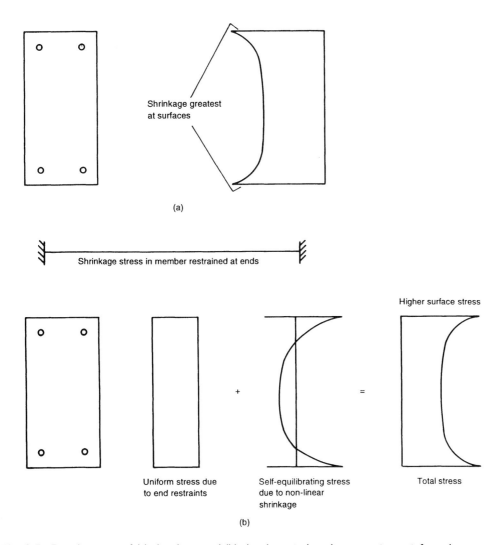

Fig. 8.5. Development of (a) shrinkage and (b) shrinkage-induced stresses in a reinforced-concrete member

deformation of the surface concrete will be restrained by the interior concrete, and higher tensions will be developed near the surface. This is illustrated in Fig. 8.5. The higher tensile stresses at the surface will lead to cracking occurring at a lower load than would be predicted on the basis of a linear distribution of stress across the section. Since the cracking load will be smaller than predicted, a smaller amount of reinforcement will be necessary to ensure that controlled cracking occurs. The function of the factor k is thus to reduce the minimum steel area in cases where such non-linear stress distributions can occur.

Finally, it is necessary to consider what value should be chosen for the tensile strength of the concrete, f_{ct}. It will be seen from the equations above that the minimum area of reinforcement is proportional to the tensile strength of the concrete. It thus seems logical to take as the tensile strength of the concrete an estimate of its likely maximum value. Such a value is not easy to establish, and, furthermore, would lead to levels of reinforcement that would be considered impractically high. For this reason, a more pragmatic approach has been adopted, and the provisions of the code aim to give a minimum area which is not too different from current practice.

The above considerations lead to *equation (7.1)* given in *clause 7.3.2* for minimum reinforcement:

Clause 7.3.2

$$A_{\text{s, min}}\sigma_{\text{s}} + A_{\text{p}}\xi_1 \Delta\sigma_{\text{p}} \geq k_{\text{c}}kf_{\text{ct, eff}}A_{\text{ct}}/\sigma_{\text{s}} \tag{D8.10}$$

where

k_{c} is a coefficient taking account of the form of the loading.

k is a coefficient taking account of the possible presence of non-linear stress distributions.

$f_{\text{ct, eff}}$ is the mean value of the tensile strength of the concrete effective at the time when the cracks first form. EN 1992-1-1 suggests that this should be taken as f_{ctm}, given in *Table 3.1*, though a lower value may be used where it is clear that the cracking will occur before 28 days.

A_{ct} is the area of concrete in tension immediately before the formation of the first crack.

σ_{s} is the steel stress, which can generally be taken as the characteristic yield strength of the steel, though, as will be seen, there are cases where it may be more convenient to use other, lower values.

A_{p} is the area of prestressing tendons within 150 mm of the ordinary bonded reinforcement.

$\Delta\sigma_{\text{p}}$ is the increase in stress in the prestressing tendons above that corresponding to zero stress in the concrete at the same level.

ξ_1 is a coefficient taking account of the different bond properties of the prestressing tendons and the ordinary reinforcement.

Interpretation of the formula

EN 1992-1-1 provides a considerable amount of guidance on how to arrive at values for the coefficients k and k_{c} and the other parameters in the equation. This is summarized below:

Values for k_{c}

k_{c} takes a value of 1.0 for pure tension, 0.4 for pure flexure in rectangular reinforced concrete sections or the webs of reinforced concrete box or flanged sections. The value of k_{c} can be reduced if there is a compressive force applied to the section, either by prestress or other means. Similarly, k_{c} will be increased where there is an applied tension.

Where, under the characteristic combination of loads, the stress in the section remains more compressive that $f_{\text{ct, eff}}$ and k_{c} may be taken as zero, and no minimum reinforcement is required. In other circumstances, values for k_{c} may be calculated for rectangular sections and the webs of flanged or box sections from

$$k_{\text{c}} = 0.4\{1 - \sigma_{\text{c}}/[k_1(h/h^*)f_{\text{ct, eff}}]\} \leq 1.0 \tag{D8.11}$$

where

σ_{c} is the average stress in the concrete due to external forces (i.e. if the external force is N, $\sigma_{\text{c}} = N/A_{\text{c}}$)

h^* $= h$ (the overall depth of the section in metres), where $h < 1$ m or 1.0 m where $h \geq 1.0$ m.

k_1 is a coefficient taking account of the nature of the applied external force. $k_1 = 1.5$ if the applied force is compressive or $2h^*/3h$ if it is tensile.

For a flange in tension of a flanged beam or box section, k_{c} may be calculated from

$$k_{\text{c}} = 0.9F_{\text{cr}}/A_{\text{ct}}f_{\text{ct, eff}} \geq 0.5 \tag{D8.12}$$

where

F_{cr} is the absolute value of the tensile force in the flange immediately prior to the formation of the first crack. The cracking load is calculated on the basis of $f_{\text{ct, eff}}$.

A_{ct} is the area of the flange considered.

Values for k

The basic problem here is to establish whether the possible imposed deformations are imposed from outside the element in question or whether they arise from shortening of the member itself. In cases where the imposed deformation is external to the element such as the settlement of a foundation, the question of non-linear distributions of strain across a section does not arise, and k must take a value of 1.0. Where the deformations are generated within the member considered, such as those caused by shrinkage or change in temperature of the member, lower values of k will be appropriate. In this case, k should be taken as 1.0 for members less than 300 mm deep or 0.65 for members of depth greater than 800 mm. Intermediate values may be interpolated.

Mixtures of prestressing tendons and ordinary reinforcement

The coefficient ξ_1 takes account of the different surface characteristics of prestressing tendons and ordinary reinforcement and also differences in diameter which will affect the relative bond performance. ξ_1 is given by

$$\xi_1 = \sqrt{(\xi\phi_s/\phi_p)} \tag{D8.13}$$

where ξ is the ratio of the bond strength of the prestressing tendons to that of the ordinary reinforcement, ϕ_s is the diameter of the ordinary reinforcement and ϕ_p is the diameter of the prestressing tendons. If there is no ordinary reinforcement, then $\xi_1 = \sqrt{\xi}$.

It will be seen that the estimation of the minimum reinforcement area has become a fairly complex procedure; possibly unnecessarily so in view of the extreme uncertainty about many aspects of the analysis. Nevertheless, it should be remembered that the areas of reinforcement provided in large areas of most building structures (particularly slabs) will be governed by the minimum reinforcement rules, and so these calculations will have a significant influence on the overall economy of the structure. Furthermore, there is at least some discernible logic behind the calculation compared with the arbitrary rules for minimum reinforcement given in previous design codes. What is possibly more surprising and disappointing is that these arbitrary rules are still maintained in *Sections 8* and *9* of EN 1992-1-1.

8.3.3. Principles of the cracking phenomena

For simplicity, the development of cracking will be described in terms of a reinforced concrete member subjected to pure axial tension.

Cracking caused by loading

If a continuously increasing tension is applied to a tension member, the first crack will form when the tensile strength of the weakest section in the member is exceeded. The formation of this crack leads to a local redistribution of stresses within the section. At the crack, all the tensile force will be transferred to the reinforcement, and the stress in the concrete immediately adjacent to the crack must clearly be zero. With increasing distance from the crack, force is transferred by bond from the reinforcement to the concrete until, at some distance, S_0, from the crack, the stress distribution within the section remains unchanged from what it was before the crack formed. This local redistribution of forces in the region of the crack is accompanied by an extension of the member. This extension, plus a minor shortening of the concrete which has been relieved of the tensile stress it was supporting, is accommodated in the crack. The crack thus opens up to a finite width immediately on its formation. The formation of the crack and the resulting extension of the member also reduces the stiffness of the member. As further load is applied, a second crack will form at the next weakest section, though it will not form within S_0 of the first crack since the stresses within this region will have been reduced by the formation of the first crack. Further increases in loading will lead to the formation of further cracks until, eventually, there is no remaining area of the member surface which is not within S_0 of a previously formed crack. The formation of each crack will lead to a reduction in the member stiffness. After all the cracks have formed, further loading will result in a widening of the existing cracks but no new cracks. Stresses in the concrete will be relieved by limited bond slip near the crack faces and by the formation of internal cracks. This process leads to further reductions in stiffness, but, clearly, the stiffness cannot reduce to below that of the bare reinforcement. Figure 8.6 illustrates the behaviour described above. The stepped nature of the load–deformation response is not normally discernible in tests, and the response is usually shown as a smooth curve.

Cracking due to imposed deformations

If the member is subjected to a steadily increasing strain rather than a steadily increasing load, the response illustrated in Fig. 8.6 is radically changed. When a crack forms, the reduction of stiffness resulting from the formation of the crack leads to a reduction in the

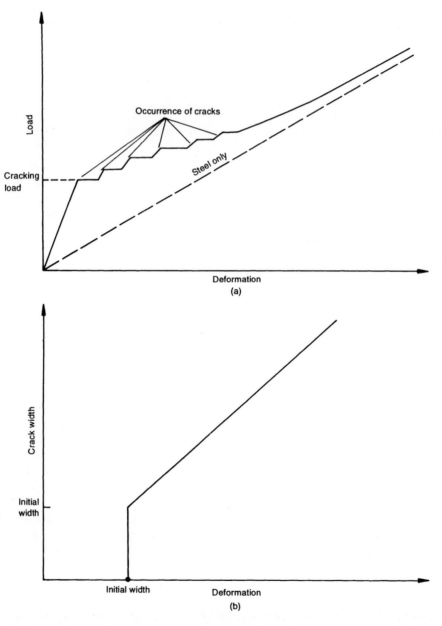

Fig. 8.6. (a) Load–deformation response of a member subjected to a steadily increasing load. (b) Crack width–deformation response in a load-controlled test

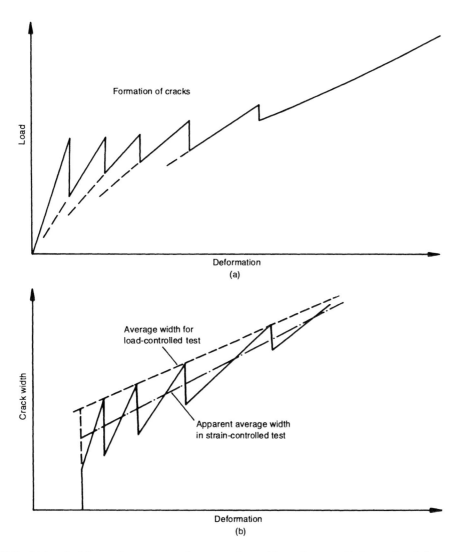

Fig. 8.7. (a) Load–deformation response for a member subjected to steadily increasing deformation. (b) Crack width–deformation response for a member subjected to steadily increasing deformation

tensile force supported by the member. This, in turn, leads to a reduction in the tensile stresses at all points along the member. Only when the strain has increased sufficiently to develop the tensile stress equal to the tensile strength of the next weakest point on the basis of the reduced stiffness will the next crack form. This results in the 'saw-toothed' load–deformation and load–crack width relationships shown in Fig. 8.7.

8.3.4. Derivation of crack prediction formulae

The development of formulae for the prediction of crack widths given in *clause 7.3.4* of EN 1992-1-1 follows from the description of the cracking phenomenon given above. If it is assumed that all the extension occurring when a crack forms is accommodated in that crack, then, when all the cracks have formed, the crack width will be given by the following relationship, which is simply a statement of compatibility:

Clause 7.3.4

$$w = S_{rm}\varepsilon_m \qquad \text{(D8.14)}$$

where w is crack width, S_{rm} is the average crack spacing and ε_m is the average strain.

The average strain can be more rigorously stated to be equal to the strain in the reinforcement, taking account of tension stiffening, ε_{sm}, less the average strain in the concrete at the surface,

ε_{cm}. Since, in design, it is a maximum width of crack which is required rather than the average, the final formula given in EN 1992-1-1 is

$$w_k = S_{r,max}(\varepsilon_{sm} - \varepsilon_{cm}) \tag{D8.15}$$

Since no crack can form within S_0 of an existing crack, this defines the minimum spacing of the cracks. The maximum spacing is $2S_0$, since if a spacing existed wider than this, a further crack could form. It follows that the average crack spacing will lie between S_0 and $2S_0$. It is frequently assumed to be $1.5S_0$.

It is in the calculation of S_{rm} that the most significant differences arise between the formulae in national codes. The distance S_0, and hence S_{rm}, depends on the rate at which stress can be transferred from the reinforcement, which is carrying all the force at a crack, to the concrete. This transfer is effected by bond stresses on the bar surface. If the bond stress is assumed to be constant along the length S_0 and that the stress will just reach the tensile strength of the concrete at a distance S_0 from a crack, then

$$\tau\pi\phi S_0 = A_c f_{ct}$$

where τ is the bond stress, A_c is the area of concrete, f_{ct} is the tensile strength of concrete and ϕ is the bar diameter.

Taking $\rho = \pi\phi^2/4A_c$ and substituting for A_c gives

$$S_0 = f_{ct}\pi\phi/4\rho$$

From this,

$$S_{rm} = 0.25k\phi/\rho$$

where k is a constant depending on the bond characteristics of the reinforcement. This is the oldest form of relationship for the prediction of crack spacings. More recent studies have shown that the cover also has a significant influence, and that a better agreement with test results is obtained from an equation of the form

$$S_{rm} = kc + 0.25k_1\phi/\rho$$

where c is the cover. This formula has been derived for members subject to pure tension. In order to be able to apply it to bending, it is necessary to introduce a further coefficient, k_2, and to define an effective reinforcement ratio, ρ_{eff}. These modifications take account of the different form of the stress distribution within the tension zone and the fact that only part of the section is in tension. k_2 and ρ_{eff} can be derived empirically from tests. The resulting formula is

$$S_{rm} = 2c + 0.25k_1k_2\phi/\rho_{eff} \tag{D8.16}$$

Here, k_1 is a coefficient taking account of the bond properties of the reinforcement. A value of 0.8 is taken for high bond bars and 1.6 for smooth bars. k_2 is a coefficient depending on the form of the stress distribution. A value of 0.5 is taken for bending and 1 for pure tension. Intermediate values can be obtained from

$$k_2 = (\varepsilon_1 + \varepsilon_2)/2\varepsilon_1 \tag{D8.17}$$

where ε_1 and ε_2 are, respectively, the greater and lesser tensile strains at the faces of the member. ρ_{eff} is the effective reinforcement ratio, where A_s is the area of tension reinforcement contained within the effective area of concrete in tension, $A_{c,eff}$. This is the area of concrete in tension surrounding the reinforcement of a depth equal to 2.5 times the distance from the tension face of the member to the centroid of the tension reinforcement. A figure in the code gives definitions for other, less typical cases.

In design, it is not the average crack width which is required but a value which is unlikely to be exceeded. EN 1992-1-1 uses the characteristic crack width, which is defined as a width with a 5% probability of exceedance. It is found experimentally that a reasonable estimate of the characteristic width is obtained if the maximum crack spacing is assumed to be 1.7 times the average value. In EN 1992-1-1, therefore, the maximum spacing is used, and this is assumed to be given by

$$S_{r,\,max} = 3.4c + 0.425k_1k_2\phi/\rho_{eff} \qquad\qquad\text{(D8.18)}$$

The other parameter in the crack width equation is the average strain, $\varepsilon_{sm} - \varepsilon_{cm}$. This is obtained from *equation (7.9)* in EN 1992-1-1, and is repeated below for convenience:

$$\varepsilon_{sm} - \varepsilon_{cm} = \sigma_s/E_s - k_t f_{ct,\,eff}(1 + \alpha_e\rho_{p,\,eff})/E_s\rho_{p,\,eff} \geq 0.6\sigma_s/E_s \qquad\text{(D8.19)}$$

where σ_s is the stress in the tension reinforcement calculated assuming a cracked section, α_e is the modular ratio (E_s/E_c), and k_t is a factor depending on the duration of the load (0.6 for short-term loads and 0.4 for long-term loads).

There remains a situation where the formula described above can lead to a significant overestimate of the likely cracking. The reason for this may be understood by considering the element shown in Fig. 8.5. This is an unreinforced element subjected to an axial load applied at an eccentricity sufficiently large for part of the section to be in tension. If the load is sufficient, the section will crack. The formation of this crack will not result in failure but merely a redistribution of stresses locally to the crack. Some distance away from this crack the stresses will remain unaffected by the crack. It is found that the crack affects the stress distribution within a distance roughly equal to the height of the crack on either side of the crack. Thus, by the same arguments used earlier, the spacing of the cracks can eventually be expected to be between the crack height and twice the crack height. This leads to the following relationship:

$$S_{rm} = h - x \qquad\qquad\text{(D8.20)}$$

where h is the overall depth of the section and x is the depth of the neutral axis. This formula applies not only to members subject to axial compression but also to any situation where the cracks, when they form, do not pass right through the section. The effect of bonded reinforcement in the section is almost always to give a calculated crack spacing much smaller than given by the above equation. The equation does, however, give a maximum, or limiting, value for the spacing, and there are a number of practical situations where it can be used with advantage. One particular case is that of a prestressed beam without any bonded normal reinforcement. The bond of prestressing tendons or wires is often much inferior to that on normal high bond reinforcement. A safe estimate of the crack spacing, and hence the crack width, can be made by treating the prestress as an external load, calculating the depth of the tension zone under the loading considered and applying the above formula. *Clause 4.4.2.4(8)* permits this procedure.

Clause 4.4.2.4(8)

8.3.5. Checking cracking without direct calculation

Flexural or tension cracking can normally be controlled by application of simplified detailing rules (*clause 7.3.3*). These take the form of tables of maximum bar diameters or maximum bar spacings. The tables (*Tables 7.2* and *7.3* in EN 1992-1-1) were derived from parameter studies carried out using the crack width calculation formulae described above. It should be noted that, for load-induced cracking, **either** *Table 7.2* **or** *Table 7.3* may be used; it is **not** necessary to satisfy both tables. For cracking due to restrained imposed deformations, *Table 7.2* should be used. For reinforced concrete, the tables can be presented graphically, as in Fig. 8.8. For load-induced cracking, the steel stress is the stress under the quasi-permanent combination of loading. This may be calculated approximately from the relation:

Clause 7.3.3

$$f_{sqp} = f_{yk}N_{qp}/1.15N_{ud} \qquad\qquad\text{(D8.21)}$$

where f_{sqp} is the steel stress under the quasi-permanent load, N_{qp} is the quasi-permanent load and N_{ud} is the design ultimate load.

For prestressed concrete beams containing bonded untensioned reinforcement, it is possible to make an estimate of the stress in the untensioned steel from Figure 8.9. This is done by calculating a fictitious value for $f_{ct,\,eff}$ equal to the tensile stress in the concrete at the

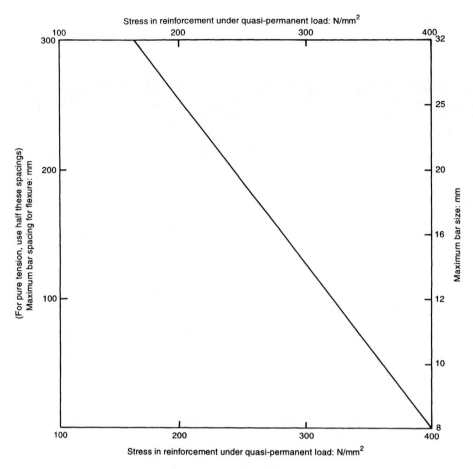

Fig. 8.8. Maximum bar sizes and spacings for crack control

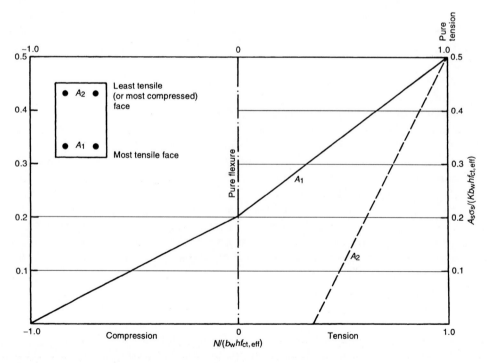

Fig. 8.9. Design chart for assessing minimum areas of reinforcement

beam soffit calculated on the assumption that the concrete is uncracked. A value of $N/bhf_{ct,eff}$ can now be calculated and a value for $\rho\sigma/f_{ct,eff}$. Substituting for ρ and $f_{ct,eff}$ will give a value for σ_s. An example will illustrate the procedure.

Example 8.2

Carry out a check for cracking for a prestressed T beam having a rib breadth of 250 mm, an overall depth of 600 mm, a flange breadth of 1000 mm and a flange depth of 180 mm. The prestress N after allowance for losses is 700 kN at an eccentricity of 280 mm and the moment under the frequent combination of loads is 330 kN m.

It can be found from standard tables that the section modulus for stresses at the bottom of the section is $0.222bh^2 = 20 \times 10^6$, and the area of the section is $285 \times 10^3 \text{ mm}^2$. The stress at the soffit of the section is thus given by

$$f_{ct,eff} = 700/285 + 700 \times 0.28/20 - 330/20 = -4.09$$

Hence

$$N/A_c f_{ct,eff} = 700/285/4.09 = 0.6$$

From Fig. 8.4c,

$$A_s\sigma_s/b_r h f_{ct,eff} = 0.08$$

Hence,

$$A_s\sigma_s = 0.08 \times 250 \times 600 \times 4.09 = 49\,080$$

Assume we choose $2 \times$ No. 12 mm bars. This gives $A_s = 226 \text{ mm}^2$, and hence $f_s = 217 \text{ N/mm}^2$. For this stress, *Table 4.11* in EN 1992-1-1 gives a maximum bar diameter of between 12 and 16 mm, so the design is satisfactory. *Table 4.12* gives a maximum bar spacing of about 130 mm, which would be difficult to achieve with only two bars. However, since *Table 4.11* has been satisfied, this does not matter.

It is, of course, possible to calculate the stress in the reinforcement rigorously on the basis of a cracked section. This would have given a significantly lower stress if proper allowance had been made for the prestressing tendons in the section, so the proposed method is on the safe side. Using the rigorously calculated stress and the formulae given for the calculation of crack width results in a width below 0.2 mm.

EN 1992-1-1 gives no way of using the tables where there is no untensioned reinforcement in the section. By interpreting the rules given in *clause 4.4.2.4(4)*, it seems a reasonable approximation to calculate a steel stress as described above. This will be a change in stress in the tendons beyond the condition where the concrete at the level of the tendons is at zero stress. To allow for the inferior bond of the tendons, the bar diameters given by *Table 4.11* or the bar spacings from *Table 4.12* should be divided by 2.5.

8.3.6. Checking cracking by direct calculation

The formula for calculating the design crack width has been discussed in Section 8.3.3, and no further discussion is considered necessary.

8.4 Control of deflections

8.4.1. General

Deflection control is dealt with in Eurocode 2 in chapter [7.4]. [7.4.2] deals with the control of deflection using simple 'deemed to satisfy' rules, and [7.4.3] deals with the calculation of deflections. It was the view of the drafting committee that, in general, the calculation of deflections gave an unwarranted impression of precision in what was a very uncertain process. It was felt that the use of simple rules, such as limits to span/effective depth ratios, was a perfectly adequate approach for all normal situations. In this chapter, however, the intentions of the code and the derivation of the simplified methods will be understood better if the calculation of deflections is covered first. In design for all service-ability limit states, there are four necessary elements. These are:

- Criteria defining the limit to satisfactory behaviour.
- Appropriate design loads.
- Appropriate design material properties.
- Means of predicting behaviour.

Each of these will now be considered in turn with reference to deflections.

8.4.2. Deflection limits

The selection of limits to deflection which will ensure that the structure will be able to fulfill its required function is a complex process and it is not possible for a code to specify simple limits which will meet all requirements and still be economical. For this reason Eurocode 2 makes it clear in [7.4.1] that it is the responsibility of the designer to agree suitable values with the client, taking into account the particular requirements of the structure. Limits are suggested in the code but these are for general guidance only; it remains the responsibility of the designer to check whether these are appropriate for the particular case considered or whether some other limits should be used.

There are two basic issues which are considered to influence the choice of limits. These are appearance and function.

Appearance is important because it is found that the occupants of structures find it upsetting if the floors appear to be sagging. Some research has been carried out on this highly subjective aspect of deflection control and it is generally accepted that sag will be unnoticeable provided the central deflection of a beam relative to its supports is less than about span/250.

Function is more difficult to cover as the range of possibilities is large. Examples of situations where deflections may lead to impairment of function are:

(i) Deflection of beams or slabs leading to cracking of partitions supported by the member considered.
(ii) Deflections causing doors to jam or windows to break.
(iii) Varying deflections leading to misalignment of apparatus or machinery mounted on the member considered.
(iv) Damage to brittle finishes.
(v) Unacceptable vibrations or an upsetting feeling of 'liveliness' in the structure.

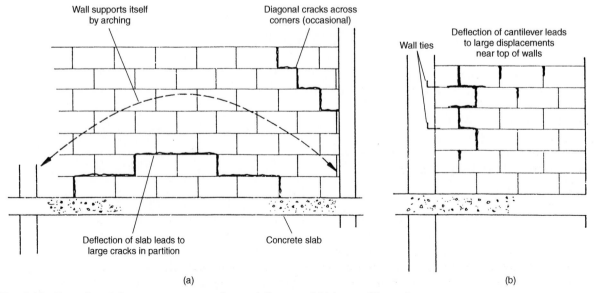

Fig. 8.10. Examples of damage to partitions due to deflection of (a) beam, (b) cantilever

Of the above list, damage to brittle partitions is probably the most common problem and the one which code limits are generally formulated to avoid. A number of typical cases of damage to partitions caused by deflections are illustrated in Figure 8.10. The width of the cracks from this cause are frequently large; 2 to 5 mm being not uncommon. It is difficult to define limits which will avoid such cracking entirely and surveys have found cracking with deflections below span/1000. Such small limits to deflection are generally considered uneconomic and many codes (including Eurocode 2) suggest a higher limit of span/500. Some codes suggest limits as high as span/300. The limit chosen in a particular case may depend on explaining to the client the consequences of various options. For example, a deflection limit of span/300 may be quite acceptable if the client is prepared to accept that significant cracking may occur in partitions after a year or so but that, if the cracks are filled up, they will very probably not reappear. This may be a considerably more economical approach than increasing the depth of the members to a level which will ensure no cracking. The limit can, in any case, be relaxed or ignored if the partitions are either flexible enough to accommodate the deflection or are detailed so that some relative movement between the partition and the supporting member can occur.

8.4.3. Design loads
When checking the overall deflection to ensure that the appearance is not impaired, it may be reasonable to take the view that, since in concrete structures the dominant part of the deflection is that occurring under the permanent loads, it is only necessary to check the deflection under the quasi-permanent combination of loads. The possibility that the deflection may occasionally exceed the calculated value is considered to be unimportant in this case.

The code is not specific about the loading that should be used when checking for impairment of function but, again, since the major part of the deflection will be due to creep and shrinkage and since the calculation of deflections is a highly approximate process, it may be reasonable to use the quasi-permanent load.

8.4.4. Material properties
Clause 7.4.3.(2)P in Eurocode 2 states that the calculation method should represent the true behaviour of the structure to an accuracy appropriate to the objectives of the calculation. This suggests that the material properties assumed in the calculation should reflect a best estimate of their values rather than a lower bound. This is confirmed by the provisions of *clause 7.4.3(4)* where the average values of tensile strength and modulus of elasticity are used for concrete (f_{ctm} and E_{cm}).

8.4.5. Model of behaviour
8.4.5.1. Short term behaviour
Figure 8.11 gives an idealised picture of the load – deformation characteristics of a reinforced concrete beam. It is convenient to consider the curve to be made up of three phases:

Phase 1 – uncracked
In this phase, the tensile strength of the concrete has not been exceeded. The section behaves elastically and its behaviour can be predicted on the basis of an uncracked section but with allowance made for the reinforcement.

Phase 2 – cracked
In this phase, the concrete has cracked in tension. The concrete in compression and the reinforcement may, however be considered to remain elastic. The behaviour of the tension zone is complex. At a crack the concrete in tension carries no stress. However, between cracks, bond transfers stress from the reinforcement to the concrete so that, with increasing distance from a crack, the tension carried by the concrete increases. The behaviour illustrated in Figure 8.11 for this phase reflects the average state, where the tension zone carries some

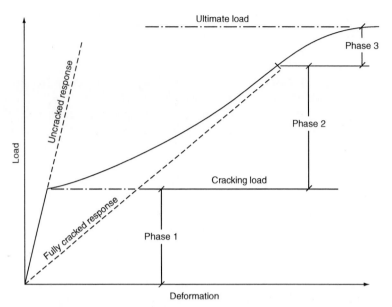

Fig. 8.11. Idealized load–deformation characteristics of a reinforced concrete member

tension. This is perfectly satisfactory for deflection calculations since the calculation of the deflection requires the integration of the behaviour over the length of the beam.

Phase 3 – inelastic

In this phase, either the steel has yielded or the concrete is stressed to a level where the assumption of elasticity ceases to be reasonable or both. This phase is generally only reached at loads well above those likely to occur in normal service and so is not of interest for serviceability calculations.

It is the behaviour in phase 2 which causes difficulties and all current design approaches are empirical and approximate.

It is possible to go some way towards defining the characteristics of a prediction method since certain limits to behaviour can be defined. These are:

(a) At the instant of cracking, when the tensile strength of the concrete is just attained, the response of the member must lie on the Phase 1 line.
(b) Since cracking effectively reduced the stiffness of the member, behaviour after cracking must lead to curvatures greater than the Phase 1 curvature.
(c) The maximum possible curvature corresponds to the condition where the concrete in tension carries absolutely no stress. This is the response that would be calculated on the basis of a cracked transformed section. The curvature corresponding to this is indicated on Figure 8.11.

In practice, experiments show that, at the cracking moment, the behaviour lies on the Phase 1 line and, as the load is increased above the cracking load, the response tends towards the fully cracked response. This is what would be expected since increase in cracking and bond slip in the region of the cracks leads to an increasing loss of effectiveness of the concrete in tension. Many formulae have been developed which satisfy the basic requirements set out above and there is probably not a lot to choose between them. The method given in Eurocode 2 is one such. It has the added practical advantage of being relatively easy to apply.

The basic concept of the method is illustrated in Figure 8.12. Considering a length of a beam bounded by two cracks, the assumption is made that some length close to the cracks is fully cracked while the remainder is assumed to be uncracked. Considering the whole crack spacing, S, it will be seen that:

the length S is considered fully cracked and

the length $(1 - \zeta)S$ is considered uncracked.

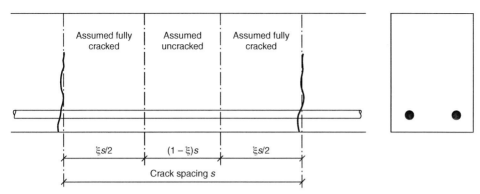

Fig. 8.12. Model of behaviour in phase 2

For the simple case of pure flexure, the rotation over the length S is given by:

$$\theta = \zeta S(1/r)_2 + (1 - \zeta)S(1/r)_1$$

Hence, the average curvature is given by:

$$(1/r)_\mathrm{m} = \theta = \zeta(1/r)_2 + (1 - \zeta)(1/r)_1$$

The same principle can be adopted for the calculation of any parameter relating to behaviour under any condition of loading. Thus, for example, the average steel strain may be calculated using the equation:

$$\varepsilon_\mathrm{sm} = \zeta\varepsilon_\mathrm{s2} + (1 - \zeta)\varepsilon_\mathrm{s1}$$

In the above equations, the subscripts 1 and 2 indicate behaviour calculated assuming the section to be uncracked and fully cracked respectively.

An alternative way of visualising the idealisation is shown in Figure 8.13 for the strain in the reinforcement.

So-far, the model is simply a convenient way of expressing the condition that the actual behaviour after cracking must lie between that calculated for an uncracked section and that calculated for a fully cracked section. Any actual result could be simulated by the suitable choice of the distribution coefficient. Empiricism now enters the process in order to define a suitable expression for the distribution coefficient. Any suitable expression must have the property that, at the cracking load, it must take the value 0 and, with increasing load, the value must approach 1.0. Eurocode 2 adopts the expression:

$$\zeta = 1 - \beta(\sigma_\mathrm{sr}/\sigma_\mathrm{s})^2$$

β is a coefficient characterising the influence of the duration of loading or of repeated loading on the average strain. It is proposed that:

for sustained loads or many cycles of repeated loading, $\beta = 0.5$
for a single short-term load, $\beta = 1.0$

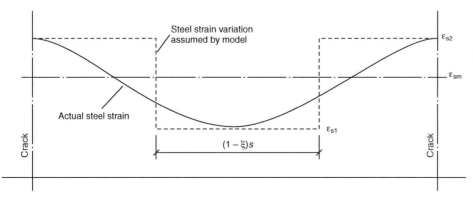

Fig. 8.13. Alternative visualization of model of behaviour in phase 2

163

σ_{sr} is the steel stress calculated on the basis of a cracked section under the loading which just causes the tensile strength of the concrete to be attained at the section considered.

σ_s is the steel stress under the loading considered calculated on the basis of a cracked section.

8.4.5.2. Long term behaviour

There are three factors to be considered in calculating the long term behaviour. These are: creep, shrinkage and, in cracked sections, reduction of the tensile stresses in the concrete in the tension zone due to the spreading with time of cracking and local bond failure. These will be considered in turn.

(a) Creep

Creep can conveniently be dealt with using the effective modulus method. This is not generally considered to be the most accurate method of modeling creep but has the advantage of simplicity. In view of the approximate nature of deflection calculations, however, a more accurate but less convenient method would not be justified. The effective modulus of elasticity of the concrete, taking account of creep is given by the relation:

$$E_{c,\text{eff}} = E_{cm}/(1 + \varphi(\infty, t_o))$$

where:

$E_{c,\text{eff}}$ is the effective modulus of elasticity of the concrete.
E_{cm} is the short term modulus of elasticity of the concrete.
φ is the creep coefficient.

Values for the creep coefficient are discussed in 8.1.2. above.

(b) Shrinkage

Uniform shrinkage of an unrestrained, unreinforced member will simply lead to an overall shortening without either curvature or stresses being induced. Reinforcement, which does not shrink, will restrain the shrinkage to some degree. This will lead to compression in the reinforcement and tension in the concrete. Where the reinforcement within the section is unsymmetrical, the restraint provided by the reinforcement will also be unsymmetrical and a curvature will be induced. In shallow members this curvature can be large enough to produce significant deflections which will need to be taken into account in any calculations.

The method given in Eurocode 2 for the calculation of shrinkage deformations can be derived for uncracked sections as follows.

Consider, for simplicity, the simply supported rectangular beam shown in Figure 8.14. If this beam is constrained to shorten by the amount of the free shrinkage, a compressive stress will be induced in the reinforcement of $\varepsilon_{cs}E_s$. This is equivalent to a force, N_{cs}, equal to $\varepsilon_{cs}E_sA_s$ where:

ε_{cs} = free shrinkage strain
E_s = modulus of elasticity of the reinforcement
A_s = area of reinforcement

If the system is now released, the beam will deform under the released force in the steel. This leads to a curvature which will be given by:

$$(1/r)_{cs} = N_{cs}e/E_cI_1 = \varepsilon_{cs}E_sA_se/E_cI_1 = \alpha_e\varepsilon_{cs}S/I_1$$

where:

α_e = modular ratio = E_s/E_c
e = eccentricity of reinforcement
I_1 = second moment of area of the uncracked section

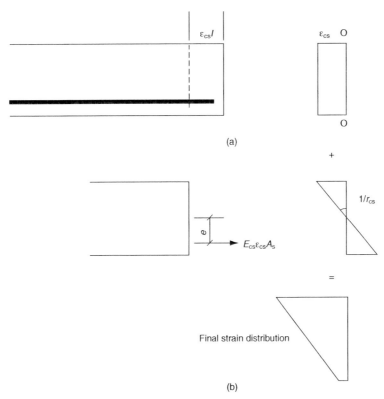

Fig. 8.14. Deformation due to shrinkage of a rectangular beam: (a) if free, concrete would shorten by $\varepsilon_{as} l$; (b) shortening of concrete would induce a force in the reinforcement and hence develop an internal moment, hence a curvature

S = first moment of area of the reinforcement about the centroid of the concrete section ($= A_s e$ for a singly reinforced section).

In the above equations, E_c should clearly be an effective value allowing for the effects of creep since shrinkage is a long term effect.

The extension of this method to cover cracked sections is speculative but seems reasonable. It is therefore suggested simply that, for cracked sections, I_2, the second moment of area of the cracked section, is substituted for I_1 and that S is also calculated for a cracked section in order to calculate the appropriate value of shrinkage curvature for the fully cracked state. The actual curvature is then calculated using the distribution coefficient approach set out in 8.4.5.1. above.

Values for free shrinkage are considered in 8.1.2. above.

8.4.5.3. Example of curvature calculation
The long term curvature will be calculated for the mid-span section of a beam the cross-sectional details of which are shown in Figure 8.15. The moment under the combination of loads considered is 65 kNm. It may be assumed that the tensile strength of the concrete, f_{ctm} is 2.2N/mm^2, the elastic modulus, E_{cm}, is 29 kN/mm^2, the creep coefficient is 2.0 and the free shrinkage strain is 300×10^{-6}.

The cracking moment is given by:

$$M_{cr} = f_{ctm} bh^2 / 6 = 2.2 \times 300 \times 500^2 / 6 \times 10^{-6} \text{ kNm}$$

$$= 27.5 \text{ kNm}$$

Since this is much less than the applied moment, the section is clearly cracked.
The effective modulus of elasticity of the section is given by:

$$E_{c,eff} = E_{cm} / (1 + \varphi(\infty, t_o)) = 29/(1 + 2) = 9.67 \text{ kN/mm}^2.$$

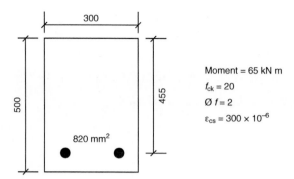

Fig. 8.15. Section used in Example 9

The curvature of the uncracked section is given sufficiently accurately as:

$$(1/r)_1 = M/EI = 65 \times 10^6 \times 12/(9.67 \times 10^3 \times 300 \times 500^3)$$

$$= 2.15 \times 10^{-6}$$

In the above calculation, no allowance has been made for the influence of the reinforcement on the second moment of area of the uncracked section. This could be taken into account but the difference to the final result will not be significant.

The effective modular ratio, $\alpha_e = E_s/E_{c,eff} = 20.68$.

This leads to a transformed reinforcement ratio of:

$$= \alpha_{eAs}/bd = 20.68 \times 820/300 \times 455 = 0.124$$

Figure 8.5 gives $x/d = 0.39$ and hence $x = 177\,\text{mm}$.

The steel stress can now be calculated as:

$$M/A_s(d - x/3) = 65 \times 10^6/(820(455 - 177/3) = 200\,\text{N/mm}^2$$

From this the curvature of the fully cracked section can be calculated as:

$$(1/r)_2 = \varepsilon_s/(d - x) = 200/200000(455 - 177) = 3.6 \times 10^{-6}$$

It is now necessary to calculate the distribution factor, ζ.

The steel stress at the cracking moment can be obtained pro-rata from the stress under the design moment as $200 \times M_{cr}/M = 200 \times 27.5/65 = 84.6\,\text{N/mm}^2$.

For long term loads, $\beta = 0.5$ hence:

$$\zeta = 1 - 0.5(84.6/200)^2 = 0.91$$

$$1/r = \zeta(1/r)_2 + (1 - \zeta)(1/r)_1$$

$$= (0.91 \times 3.6 + 0.09 \times 2.15) \times 10^{-6} = 3.47 \times 10^{-6}$$

The shrinkage curvature for a cracked and uncracked section is now calculated.

For the uncracked section, this is given by:

$$(1/r)_{cs1} = 300 \times 10^{-6} \times 20.68 \times (820 \times 205) \times 12/300 \times 500^3$$

$$= 0.33 \times 10^{-6}$$

For the cracked section, Figure 8.5 gives the second moment of area as

$0.066 \times 300 \times 500^3 = 250 \times 10^8$.

Hence:

$$(1/r)_{cs2} = 300 \times 10^{-6} \times 20.68 \times 820 \times (455 - 177)/250 \times 10^8$$

$$= 0.57 \times 10^{-6}$$

Using the same distribution factor as for calculating the curvature due to load enables the actual shrinkage curvature to be calculated as:

$$(1/r)_{cs} = (0.91 \times 57 + 0.09 \times .33) \times 10^{-6} = 0.55 \times 10^{-6}.$$

The total curvature is thus $(3.47 + 0.55) \times 10^{-6} = 4.02 \times 10^{-6}$.

It will be seen that, even with the relatively low reinforcement ratio used in this example, the calculated curvature does not differ very much from that calculated assuming a fully cracked section. For sections with higher percentages of reinforcement, it is probably sufficiently accurate to simply calculate long term curvatures on the basis of a fully cracked section.

8.4.5.4. Calculation of deflections from curvatures

The deflection of a member is calculated by double integration of the curvature over the length of the beam and introduction of appropriate boundary conditions.

The most general way of achieving this is by calculation of the curvature at intervals along the member and then calculating the deflection of numerical integration. Using this method, it is possible to take account of the complex relation between moment and curvature due to the development of cracking and also the effects of varying steel percentage or section shape. The procedure may be iterative for indeterminate beams since the moment field calculated using normal methods of analysis is unlikely to be the correct one when allowance has been made for the varying stiffness from point to point along the beam due to the development of cracking.

A calculation of this type can conveniently be set out in a tabular form and is well suited to the use of a spread sheet program. Such an approach will be illustrated in the example below.

The deflection will be calculated of a simply supported beam of span 5.1 m supporting a uniformly distributed load of 20 kN/m. The section details, material properties, the cracked neutral axis depth and second moment of area are as given in 8.4.5.3 above. The moment at any section is given by the relation:

$$M = nLx/2 - nx^2/2$$

As in the previous example, the mid-span moment can be calculated to be 65kNm. The curvatures at tenth points along the span are calculated using the method illustrated in 8.4.5.3. above by the spreadsheet as shown in Table 8.3 below.

It will be seen that the values in the table for mid-span are the same as in the example in 8.4.5.3. Table 8.4 continues the calculation by carrying out the integration and introducing the boundary conditions. The integrations are carried out using the trapezoidal rule. The first integration, giving the uncorrected rotations at each section is thus given by:

$$\theta_i = \theta_{(i-1)} + ((1/r)_i + (1/r)_{(i-1)})/2 \times L/10$$

Table 8.3. Calculation of curvatures at tenth points

Fraction of span	Moment kNm	$1/r_1 \times 10^6$	$1/r_2 \times 10^6$	Factor	$1/r_{cs} \times 10^6$	$1/r_{(tot)} \times 10^6$
0	0.0	0.00	0.00	0.00	0.33	0.33
0.1	23.4	0.77	1.30	0.31	0.40	1.34
0.2	41.6	1.38	2.31	0.78	0.52	2.62
0.3	54.6	1.81	3.03	0.87	0.54	3.41
0.4	62.4	2.07	3.46	0.90	0.55	3.87
0.5	65.0	2.15	3.60	0.91	0.55	4.02
0.6	62.4	2.07	3.46	0.90	0.55	3.87
0.7	54.6	1.81	3.03	0.87	0.54	3.41
0.8	41.6	1.38	2.31	0.78	0.52	2.62
0.9	23.4	0.77	1.30	0.31	0.40	1.34
1	0.0	0.00	0.00	0.00	0.33	0.33

Table 8.4. Calculation of deflection from curvatures

Fraction of span	$1/r_{(tot)} \times 10^6$	Uncorrected rotation	Uncorrected deflection	Correction	Final deflection
0	0.33	0.0	0.0	0.0	0.0
0.1	1.34	0.4	0.1	3.5	−3.4
0.2	2.62	1.4	0.6	7.0	−6.4
0.3	3.41	3.0	1.7	10.5	−8.8
0.4	3.87	4.8	3.7	13.9	−10.3
0.5	4.02	6.8	6.7	17.4	−10.8
0.6	3.87	8.9	10.7	20.9	−10.2
0.7	3.41	10.7	15.7	24.4	−8.7
0.8	2.62	12.2	21.5	27.9	−6.4
0.9	1.34	13.3	28.0	31.4	−3.4
1	0.00	13.6	34.9	34.9	0.0

Having established the uncorrected rotations, the uncorrected deflections are obtained by the second integration given by:

$$a_u = a_{u(i-1)} + (\theta_i - \theta_{(i-1)})/2 \times L/10$$

The required boundary condition is that the deflection should be zero at both supports. This can be introduced by applying a linear transformation to the uncorrected deflections. In this case, there is an uncorrected deflection of 34.88 at the right-hand end. The correction is effected by subtracting $34.88xL$ from each uncorrected deflection where x is the distance from the left-hand support. This gives the final deflected shape of the beam. This may be seen in the final column of Table 8.4.

8.4.5.5. Simplified methods of deflection calculation

The use of numerical integration is tedious and will be unnecessary in most practical situations. Simpler approaches will normally be adequate.

A simplification that may normally be made is to calculate the curvature at only one point, usually the point of maximum moment, and then assume that the shape of the curvature diagram is the same as the shape of the bending moment diagram. The deflection may then be calculated from the equation:

$$a = kL^2(1/r)$$

where k is a constant which depends upon the shape of the bending moment diagram. Values for k are given in Table 8.5, taken from the UK code, BS8110 Part2:1985.

For a uniformly distributed load on a simply supported beam, k is 0.104. Hence the deflection of the beam used in the spreadsheet example above may be calculated as:

$$a = 0.104 \times 5.1^2 \times 4.019 = 10.87 \, \text{mm}$$

This will be seen to be very close to the value of 10.77 mm calculated by the more rigorous approach. Errors will be greater where the maximum moment is closer to the cracking moment but will not generally exceed 10%.

It is suggested that, where the deflection of a continuous beam is being calculated, the critical section at which the curvature is calculated should be the section with the maximum sagging moment, even though the hogging moment over the support may be bigger.

An alternative simplified approach is to calculate the deflections on the basis of the uncracked and the fully cracked section and then use the distribution coefficient to combine them. The result would have been the same, though the shrinkage curvatures would have had to be converted to deflections.

Table 8.5. Values of k for various bending moment diagrams

Loading	Bending moment diagram	k
M ⟲ — — ⟳ M	M	0.125
al, W, l	$M = Wa(1-a)/l$	$\dfrac{3-4a^2}{48(1-a)}$ If $a = \dfrac{1}{2}$, $k = \dfrac{1}{12}$
• — — ⟳	M	0.0625
al, $W/2$, $W/2$, al	$M = (Wal)/2$	$0.125 = \dfrac{a^2}{6}$
q	$(ql^2)/1/8$	0.104
q	$(ql^2)/15.6$	0.102
q	M_A, M_C, M_B	$k + 0.104\left(1 - \dfrac{\beta}{10}\right)$ $\beta = \dfrac{M_A + M_B}{M_C}$
al, W	Wal	End deflection $= \dfrac{a(3-a)}{12}$ Load at end $k = 0.333$
al, q	$(qa^2l^2)/2$	$\dfrac{a(4-a)}{12}$ If $a = l$, $k = 0.25$
	M_A, M_C, M_B	$k = 0.083\left(1 - \dfrac{\beta}{4}\right)$ $\beta = \dfrac{M_A + M_B}{M_C}$
al, al	$[(Wl^2)/24]\,(3-4a^2)$	$\dfrac{1}{80}\dfrac{(5-4a^2)^2}{3-4a^2}$

8.4.5.5. Accuracy of deflection calculations

There are many sources of uncertainty which will influence the reliability of deflection calculations. Some of these are:

- Actual level of loading relative to design loading
- variability of tensile strength of concrete
- variability in elastic modulus of concrete
- variability of creep and shrinkage
- behaviour of a cracked tension zone
- stiffening effect of non-structural elements and finishes
- temperature effects
- age when loading is applied and load history.

Many of these factors can have very large effects on the resulting actual deflection. The result is that, while in laboratory where all the above effects are known or are measurable, an accuracy of around 20% may be achievable; the accuracy in practice is likely to be far less.

A factor of particular significance in the assessment of the deflection of slabs is the possible variation in tensile strength. This arises because the design load for deflections is commonly close to the cracking moment. The situation is shown schematically in Figure 8.16. This indicates the range of possible load – deflection curves that might be expected to occur as a result of variations in the cracking moment. It will be seen that this effect can lead to a very wide dispersion in the possible deflections. Other factors can have an equally dramatic effect on the precision of calculations.

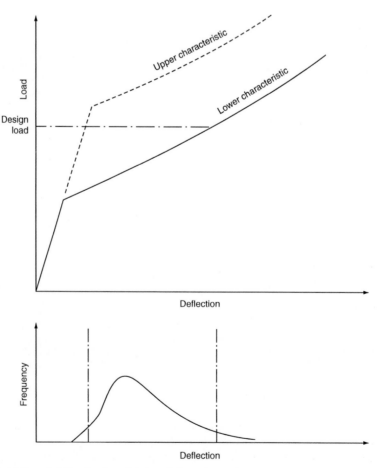

Fig. 8.16. Variability of deflection in a lightly reinforced member

Because of this wide dispersion, it may well be worth carrying out analyses to establish the possible range of deflections likely to occur rather than just a single determination in order to gain some idea of the reliability of the calculation.

8.4.6. Simplified approach to checking deflections
8.4.6.1. Basis of span/effective depth ratio approach
As has been seen, the calculation of deflections is a tedious operation and the results will be of limited reliability. Most codes therefore do not require explicit calculations to be carried out to check deflections except in special cases. The commonest method for controlling deflections is to use limiting ratios of span to effective depth and this is the approach adopted by Eurocode 2.

The basic logic of the use of span/depth ratios to control deflections is straightforward. Consider a simply supported elastic beam supporting a uniformly distributed load of n per unit length and assume that the maximum permissible stress in the material is f. The maximum moment that the critical section can withstand is thus given by:

$$M = fz = nL^2/8$$

In this equation, z is the section modulus which, for a rectangular section is $bh^2/6$.

The deflection of the beam is given by:

$$a = 5nL^4/384EI$$

Substituting for n from the previous equation gives:

$$a = 40fzL^4/384EIL^2$$

Rearranging and substituting βh for I/z gives:

$$a/L = 40f/384\beta E(Lh)$$

Since, for a given section shape and material, $40f/384\beta E$ is a constant, this can be rewritten as:

$$(a/L) = k(L/h)$$

Thus, for an elastic material, limiting the span/depth ratio will limit the ratio of the deflection to the span. This is an entirely rigorous way of controlling the deflection in these circumstances provided that the limits to deflection are expressed as fractions of the span. Reinforced concrete does not strictly fit the assumptions on which this analysis is based. However the differences are not as large as might at first appear. There are two problems to consider: the permissible stress, f and the basic stiffness properties of the section, which may be considered to be equivalent to βE. Since the use of span/depth ratios is limited to consideration of beams and slabs where the tension reinforcement will yield at ultimate, this effectively defines a limiting steel stress under service conditions which can be considered to be equivalent to f. Since the stress limit is applied at the level of the reinforcement rather than at the extreme fibre of the beam, it is more logical to use ratios of span to effective depth rather than span to overall depth. Rules can either be drafted in terms of the steel stress under service conditions or in terms of the characteristic strength, effectively assuming a constant relationship between ultimate and service stress. The second of these options is the simpler for the user and is the approach adopted in Eurocode 2. The question of the stiffness properties is more complex but can be dealt with relatively easily by assuming that, for a given section shape, βE will vary as a function of the reinforcement ratio.

One way of presenting such a system is to provide a set of basic span/effective depth ratios as a function of support conditions (ie simply supported, continuous, cantilever etc.) and a table of modification factors which are a function of steel percentage and steel characteristic strength. A further factor will be needed to correct for section shape. This is the approach which has been used in UK codes for the last 20 years.

In order to develop suitable modification factors for use in Eurocode 2, an extensive parameter study was carried out. Deflections were calculated using the calculation method

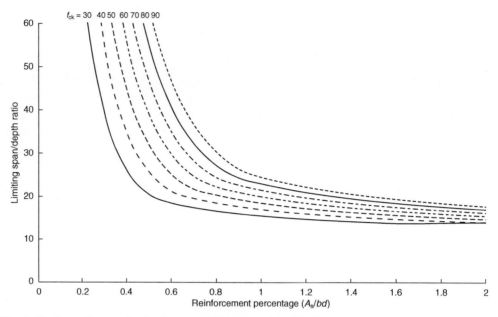

Fig. 8.17. Span-effective depth ratios for different concrete grades

described above for an extensive range of material properties, section geometries and ratios of permanent loads to characteristic loads. For each combination of variables a relationship was obtained between span/effective depth ratio and reinforcement ratio.

The result of this exercise can be presented in the form of the set of curves shown in Figure 8.17 combined with the values of k given in Table 7.4N in Eurocode 2. Figure 8.17 is drawn for $k = 1$ and therefore, to obtain the appropriate span/effective depth ratio for the particular situation considered, the value obtained from the Figure needs to be multiplied by k.

In addition, the values should be multiplied by 0.8 for flanged beams where the ratio of the flange width to the rib width exceeds 3. For values of this ratio between 1 and 3, linear interpolation may be used between 1.0 and 0.8.

8.4.6.2. Example

Check whether the internal panel of a slab with a span of 6 m, an effective depth of 200 mm and a reinforcement percentage of 0.5% will be satisfactory from the point of view of deflections. The characteristic strength of the reinforcement is 500 N/mm^2 and that of the concrete is 40 N/mm^2.

Slabs may generally be considered to have a reinforcement percentage of 0.5% or less so the right-hand column in the table may generally be used. In this example, it is stated that the steel percentage is 0.5%. From Table 7.4N in Eurocode 2, the permissible span/effective depth is 30. The actual span/effective depth ratio is $6000/200 = 30$. This is the same as the permissible value therefore the slab is OK.

Figure 8.17 could alternatively be used by assuming that the stress in the reinforcement is 310 N/mm^2 (this value is given in the note above Table 7.4N in Eurocode 2. It will be seen that Figure 8.17 gives a factor of 25 for 0.5% of tension steel and this stress while Table 7.4N gives a value for k of 1.5. The permissible span/effective depth ratio is thus $1.5 \times 25 = 37.5$. The increase in the permissible span/effective depth ratio can be seen from Figure 8.17 to be due to the use of C40 concrete where Table 7.4N has been drawn for C30 concrete.

CHAPTER 9

Durability

9.1. General

The Construction Products Directive defines certain essential requirements relating to fitness for purpose, mechanical resistance and stability and various other factors. These requirements 'must, subject to normal maintenance, be satisfied for an economically reasonable working life'. A structure having adequate durability is one which satisfies this last stipulation. This chapter is concerned with the rules given in EN 1992-1-1 and associated documents to ensure that adequate durability is achieved.

Design for durability is not covered fully by EN 1992-1-1, and reference needs to be made to a number of other standards. These are:

- BS EN 206-1 (*Concrete*, Part 1: *Specification, Performance, Production and Conformity*).
- BS 8500-1 (*Concrete – Complementary British Standard to BS EN 206-1*, Part 1:2002, *Method of Specifying and Guidance for the Specifier*).
- BS 8500-2 (*Concrete – Complementary British Standard to BS EN 206-1*, Part 2:2002, *Specification for Constituent Materials and Concrete*).
- ISO 9690 (*Production and Control of Concrete. Classification of Chemically Aggressive Environmental Conditions Affecting Concrete*).

In fact, BS EN 206 is not, in itself, a very helpful document for the designer or specifier and all the information needed for specification of concrete in the UK is included in the complementary British standard, BS 8500.

In EN 1992-1-1, durability is mainly covered in *Section 4* ('*Durability and cover to reinforcement*'), but parts of *clauses 7.2* (on stresses) and *7.3* (on cracking) are also relevant.

Clause 7.2
Clause 7.3

A good source of general information on durability is *CEB Design Guide on Durable Concrete Structures*.[13]

9.1.1. Historical perspective

Up until about the mid-1970s, durability was not seen as a serious issue for concrete. While all codes gave minimum covers for protection of the reinforcement, it was rarely felt necessary to do more than this. For example, inspection of the contents list of the CEB–FIP 1978 Model Code,[7] which formed the original base document for the drafting of EN 1992-1-1, will reveal no explicit mention of durability. During the 1970s, however, durability problems arose in very many countries. This has resulted in a complete change in attitude to the design and construction of concrete structures. A few of the more notable of these problems are:

- Very serious deterioration of bridge decks in the USA due to corrosion of the reinforcement contaminated by de-icing salts.

- Major deterioration problems in the Middle East due to chloride-induced corrosion in a particularly aggressive environment.
- Major problems in the UK in the 1970s due to reductions in strength of high-alumina cement with time. More recently, this has also become a major problem in Spain.
- Severe cracking in structures in many countries resulting from the alkali–silica reaction. While the number of affected structures may not have been great, the problem generated a large amount of publicity.
- Major problems resulting from corrosion initiated by de-icing salts on bridges in almost all countries where these are used in significant quantities in winter.

As a result of these problems, durability and, in particular, corrosion of reinforcement was probably the major area of research in the structural field during the 1980s. Inevitably, the drafters of codes of practice also found it essential to treat the subject much more fully. The Eurocodes have followed this trend.

9.1.2. Common mechanisms leading to the deterioration of concrete structures

This section lists the main deterioration mechanisms which may need to be considered in design, gives a brief description of the phenomena involved and the possible methods for dealing with them.

Corrosion of reinforcement or prestressing tendons

In normal circumstances, the highly alkaline nature of concrete protects steel embedded within it. Except under the circumstances discussed below, the pH value of the pore solution in concrete is in the region of 12–14. Steel will not generally corrode in uncontaminated concrete until the pH drops below 10. The protection is afforded by the formation of a very thin, coherent layer of iron oxide over the surface of the bar under alkaline conditions. Steel protected in this way is described as being in a passive state. Two mechanisms can lead to the destruction of this passive state. These are:

(1) Carbonation of the concrete. This is a reaction between carbon dioxide in the atmosphere and the alkalis in the cement matrix. This process starts at the surface, and, with time, penetrates slowly into the concrete. The rate of penetration of carbonation into the concrete depends upon the environment and the quality of the concrete. The rate is fastest where the relative humidity is in the range 50–60%. It is slower at higher humidities, being effectively zero at 100%. Good-quality concrete carbonates more slowly than poor quality material. The speed of the phenomenon depends on the rate at which carbon dioxide can diffuse into the concrete. This will decrease with a decrease in the water/cement ratio and hence with an increase in strength. The effect of carbonation is to reduce the alkalinity of the concrete to a level where the natural protection is lost, and corrosion may then occur if the concrete immediately surrounding the reinforcement is carbonated.
(2) The presence of chlorides in the concrete. Chlorides have the capacity to destroy the passivity of steel even where the alkalinity remains high. This usually occurs locally, giving rise to 'pitting corrosion'. Chlorides may get into the concrete from various sources, but the commonest are seawater in marine environments, de-icing chemicals on roads and additives such as calcium chloride, which was used extensively in the past as an accelerator. The rate at which chlorides penetrate into concrete depends upon the rate of application of chlorides to the concrete surface and, as with carbonation, on the quality of the concrete.

Once the passivity of the steel has been destroyed, corrosion can occur if there is (1) sufficient moisture and (2) sufficient oxygen. It is found that these two requirements can act against each other since, if the concrete is wet, oxygen cannot penetrate and, if it is dry so that there is plentiful oxygen, there is insufficient moisture for the reaction to progress. As a result, the greatest risk of corrosion is in members subjected to wetting and drying.

The normal way to design against corrosion is to ensure that there is an adequate cover to the reinforcement and that the concrete in the cover region is of a high quality and is well cured. In particularly aggressive environments, however, there are other, more expensive measures which may be taken. Possibilities are:

- Use reinforcement coated with epoxy or similar. Over recent years, this approach seems to have been somewhat discredited due to some high-profile problem cases.
- Use stainless steel reinforcement.
- Apply surface coatings to the concrete to inhibit the ingress of chlorides or carbon dioxide. Such coatings would have to be meticulously maintained to be successful for long periods of time.
- Apply cathodic protection to the structure.

One major factor in the avoidance of corrosion problems is the form of the structure. Areas of exposed concrete on which water can stand or can drain across are particularly at risk.

Frost attack

If saturated concrete is subjected to frequent freezing and thawing, the expansive effects of ice formation will disrupt the concrete. The usual manifestations of frost damage are surface spalling or the formation of systems of closely spaced surface cracks. These cracks can be seen as the precursors of spalling. Concrete which is not close to being saturated is not at risk from frost as the expansion that occurs on freezing can be accommodated in the non-water-filled pores. Frost damage can be avoided by the following methods:

- Protecting the concrete from saturation.
- Using an air-entrained concrete mix. The small bubbles within the matrix can provide pressure relief.
- Using high-strength concrete. Concrete with a strength of 45 N/mm^2 or more is generally immune to frost damage.

Alkali–aggregate reactions

There are two basic forms of reaction which occasionally occur and can damage concrete: the alkali–silica reaction and the alkali–carbonate reaction. The alkali–silica reaction is the more common. It is a reaction between the alkalis in the cement and certain forms of silica which results in the formation of a hygroscopic silica gel. This gel takes up water and expands, causing cracking. Identification of the alkali–silica reaction as the cause of damage is not straightforward. The first stage is to establish that silica gel has formed within the matrix. This can only be done by taking thin sections of the concrete and examining them under a microscope to establish the existence of gel-filled microcracks. Establishing the presence of gel is not sufficient proof in itself that the alkali–silica reaction is the cause of the observed damage as limited amounts of gel can frequently be present in concrete without any deleterious consequences. There needs to be evidence of large quantities of gel. Externally, the effects of these reactions are the formation of cracks, which may be large (several millimetres wide is not uncommon). In relatively unstressed and unreinforced concrete, these cracks can form a random 'map' pattern. In other cases they will tend to form parallel to the direction of compressive stress or reinforcement. Though large, the cracks are usually not deep, only extending 50–70 mm into the section. Their effect on structural performance is not as great as might be imagined from looking at the cracks. A reduction in the compressive and tensile strength of the concrete occurs, but this is commonly not more than about 20–30%.

The alkali–silica reaction can be avoided by three methods:

(1) Using aggregates which experience has shown to perform satisfactorily. No fully adequate test has yet been devised for assessing the potential reactivity of aggregates, so experience remains the only reliable method.
(2) Using cement with a low alkali content. The level which will avoid problems is likely to vary from country to country, depending on the geology. In the UK, a limit of 3 kg/m^3 of sodium oxide equivalent is considered adequate.
(3) Inhibiting the ingress of water.

Attack from sulphates

In the presence of water, sulphate ions can react with the tricalcium aluminate component of the matrix. This reaction causes expansion, leading to cracking and eventual disintegration of the concrete. The commonest source of sulphates is in the earth surrounding foundations but other sources are sometimes significant. Seawater contains significant amounts of sulphate but the presence of chlorides renders it harmless.

If it is established that there is a risk of sulphate attack, the possible measures to avoid it are:

• the use of sulphate-resisting cement (Portland cement having a low tricalcium aluminate content)
• the use of blended cements incorporating ground granulated blastfurnace slag
• for medium levels of risk, the use of additions such as fly ash or pozzolans are effective
• in particularly aggressive situations, it may be necessary to provide an impervious coating to the concrete.

Acid attack

Acids attack the calcium compounds in concrete, converting them to soluble salts, which can then leach away. The effect of acids is therefore to eat away, or render weak and permeable, the surface of the concrete. Very substantial amounts of acid are required to do serious damage to concrete. Acid rain, for example, will do no more than etch the surface of concrete over any reasonable design life. If the concrete is likely to be exposed to major amounts of acid, for example from some industrial process, the only way to avoid damage is to provide an impermeable coating to the concrete.

Leaching by soft water

This process is in effect a mild version of acid attack. Calcium compounds (e.g. calcium carbonate and calcium hydroxide) are weakly soluble in soft water, and can be leached out if the concrete is constantly exposed to running soft water. The process is very slow, but, in time, can lead to exposed concrete surfaces taking on an 'exposed aggregate' appearance.

Abrasion

Abrasion of concrete surfaces may occur due to trafficking of the concrete or due to sand or gravel suspended in turbulent water.

Resistance to abrasion can be obtained by using higher-strength concrete and abrasion-resistant aggregates. Resistance is also markedly improved by good curing of surfaces likely to be exposed to abrasive action.

9.1.3. Relative importance of deterioration mechanisms

Clearly, the relative importance of the various mechanisms will vary from country to country and even region to region, and no generally applicable ordering of the mechanisms can be made. However, there seems no doubt that the commonest and most serious form of degradation worldwide is corrosion of reinforcement. It can also be stated that, of the two initiating mechanisms for corrosion, carbonation and chlorides, chlorides have led to the greater amount of damage by far.

9.2. Design for durability

There are two basic steps in designing for durability. These are:

(1) Establish the aggressivity of the environment to which the member is exposed. This is analogous to establishing the design loading where the ultimate or serviceability limit states are being considered.
(2) Select materials and design the structure to be able to resist the environment for a reasonable lifetime.

The aggressivity of the environment should, in principle, be defined separately for each degredation mechanism, since the factors acting to promote one form of degredation are not necessarily the same as those promoting another. EN 1992-1-1 does this in *Table 4.1*. This classifies environments into six basic classes. These are:

(1) no risk of corrosion or attack
(2) corrosion induced by carbonation
(3) corrosion induced by chlorides
(4) corrosion induced by chlorides from sea water
(5) freeze–thaw attack
(6) chemical attack.

Each of these basic classifications, which cover the common forms of attack, are subdivided into subclasses which define the severity of the risk of attack. For example, Class 3, 'corrosion induced by chlorides', is divided into three subclasses of increasing severity: XD1, 'moderate humidity'; XD2, 'wet, rarely dry'; and XD3, 'cyclic wet and dry'. Each of the environment classes are defined by examples.

Durability is, however, difficult to define on a uniform basis for all countries which might use EN 1992-1-1, and therefore the durability provisions are all Nationally Defined Parameters. The Nationally Defined Parameters for durability in the UK are set out in the UK National Annex, and are reproduced here in Tables 9.1 and 9.2 (overleaf) for convenience. Table 9.1 gives provisions for a 50 year design life while Table 9.2 gives values for a 100 year design life.

Measures relating to the selection of materials and design of a suitable concrete mix are most conveniently found in BS 8500, while measures relating to the design of the structure, such as cover to the reinforcement, are included in EN 1992-1-1. The design of a durable structure thus requires attention to the rules in both documents

For the selection of the quality of concrete subject to freeze–thaw and concrete in aggressive ground conditions, reference should be made to Annex A of BS 8500-1.

Table 9.1. Recommendations for normal-weight concrete quality for selected exposure classes and cover to reinforcement for a 50 year intended working life and 20 mm maximum aggregate size

Exposure class	Exposure conditions	Cement/ combination types*	Equivalent designated concrete: strength class, maximum water/cement ratio, minimum cement or combination content (kg/m³)							
			Nominal cover to reinforcement including prestressing steel:							
			$15 + \Delta c$	$20 + \Delta c$	$25 + \Delta c$	$30 + \Delta c$	$35 + \Delta c$	$40 + \Delta c$	$45 + \Delta c$	$50 + \Delta c$
Carbonation-induced corrosion										
XC2	Wet, rarely dry	All	—	—	C25/30, 0.65, 260 or RC30	≪	≪	≪	≪	≪
XC3	Moderate humidity	All except IVB	—	C40/50 0.45, 340 or RC50	C32/40 0.55, 300 or RC40	C28/35 0.60, 280 or RC35	C25/30 0.65, 260 or RC30	≪	≪	≪
XC4	Cyclic wet and dry									
Chloride-induced corrosion excluding chlorides from seawater										
XD1	Moderate humidity	All	—	—	C40/50, 0.45, 360	≪	≪	≪	≪	≪
XD2	Wet, rarely dry	I, IIA, IIB-S, SRPC	—	—	—	C40/50 0.40, 380	C32/40 0.50, 340	C28/35 0.55, 320	≪	≪
		IIB-V, IIIA	—	—	—	C35/45 0.40, 380	C28/35 0.50, 340	C25/30 0.55, 320	≪	≪
		IIIB, IVB	—	—	—	C32/40 0.40, 380	C25/30 0.50, 340	C20/25 0.55, 320	≪	≪
XD3	Cyclic wet and dry	I, IIA, IIB-S, SRPC	—	—	—	—	—	C45/55 0.35, 380	C40/50 0.40, 380	C35/45 0.45, 360
		IIB-V, IIIA	—	—	—	—	—	C35/45 0.40, 380	C32/40 0.45, 360	C28/35 0.50, 340

Exposure class / Cement or combination types								
IIIB, IVB	—	—	—	—	—	C32/40 0.40, 380	C28/35 0.45, 360	C25/30 0.50, 340
Seawater-induced corrosion								
XS1 Airborne salts but no direct contact — I, IIA, IIB-S, SRPC	—	—	—	C50/60 0.35, 380	C40/50 0.45, 360	C35/45 0.50, 340	<<	<<
IIB-V, IIIA	—	—	—	C45/55 0.35, 380	C35/45 0.45, 360	C32/40 0.50, 340	<<	<<
IIIB, IVB	—	—	—	C35/45 0.40, 380	C28/35 0.50, 340	C25/30 0.55, 320	<<	<<
XS2 Wet, rarely dry — I, IIA, IIB-S, SRPC	—	—	—	C40/50 0.40, 380	C32/40 0.50, 340	C28/35 0.55, 320	<<	<<
IIB-V, IIIA	—	—	—	C35/45 0.40, 380	C28/35 0.50, 340	C25/30 0.55, 320	<<	<<
IIIB, IVB	—	—	—	C32/40 0.40, 380	C25/30 0.50, 340	C20/25 0.55, 320	<<	<<
XS3 Tidal, splash and spray zones — I, IIA, IIB-S, SRPC	—	—	—	—	—	—	C45/55 0.35, 380	C40/50 0.40, 380
IIB-V, IIIA	—	—	—	—	—	C35/45 0.40, 380	C32/40 0.45, 360	C28/35 0.50, 340
IIIB, IVB	—	—	—	—	—	C32/40 0.40, 380	C28/35 0.45, 360	C25/30 0.50, 340

*See BS 8500-2:2002, Table 1.

<< indicates that the quality of concrete given in the cell to the left should not be reduced.

Table 9.2. Recommendations for normal-weight concrete quality for selected exposure classes and cover to reinforcement for a 100 year intended working life and 20 mm maximum aggregate size

Exposure class	Exposure conditions	Cement/ combination types*	Equivalent designated concrete: strength class, maximum water/cement ratio, minimum cement or combination content (kg/m³)							
			Nominal cover to reinforcement including prestressing steel:							
			$15 + \Delta c$	$20 + \Delta c$	$25 + \Delta c$	$30 + \Delta c$	$35 + \Delta c$	$40 + \Delta c$	$45 + \Delta c$	$50 + \Delta c$
Carbonation-induced corrosion										
XC1	Dry of permanently wet	All	C20/25, 0.70, 240 or RC25	<<	<<	<<	<<	<<	<<	<<
XC2	Wet, rarely dry	All	—	—	C25/30, 0.65, 260 or RC30	<<	<<	<<	<<	<<
XC3	Moderate humidity	All except IVB	—	—	—	C40/50, 0.45, 340 or RC50	C35/45, 0.50, 320 or RC45	C32/40, 0.55, 300 or RC40	C28/35, 0.60, 280 or RC35	<<
XC4	Cyclic wet and dry									

*See BS 8500-2:2002, Table 1.

<< indicates that the quality of concrete given in the cell to the left should not be reduced.

CHAPTER 10

Detailing

10.1. General

The main guidance on detailing is contained in *Sections 8* and *9* of EN 1992-1-1. Cover requirements are in *Section 4*, under durability.

The main aspects of detailing are concerned with anchoring bars carrying stress and lapping of reinforcement. Most of the provisions regarding anchorage and laps were reputedly deduced from tests carried out in many countries, principally in Austria, France, Germany, Sweden and the USA. It is understood that the global safety factor $\gamma_m\gamma_f$ against the 5% fractile of test results is 2.1. As $\gamma_m = 1.5$ for concrete, this gives a value of $\gamma_f = 1.4$ for loads, which is clearly considered acceptable in EN 1992-1-1.

Anchorage and lap requirements are checked at the ultimate limit stage. The general detailing provisions are deemed to ensure satisfactory behaviour of the structure at serviceability conditions.

The code covers (1) high-bond bars, (2) welded mesh and (3) different types of structural members. Bars may be used in bundles with certain limitations.

In the main, the detailing requirements are governed by bond-related phenomena, which are significantly influenced by:

- the surface characteristics of the bars (plain, ribbed)
- the shape of the bars (straight, with hooks or bends)
- the presence of welded transverse bars
- the confinement offered by concrete (mainly controlled by the size of the concrete cover in relation to the bar diameter)
- the confinement offered by non-welded transverse reinforcement (such as links)
- the confinement offered by transverse pressure.

The rules governing detailing make allowance for the above. Particular emphasis is placed on the need for adequate concrete cover and transverse reinforcement to cater for tensile stresses in concrete in regions of high bond stresses.

Bond stresses for high-bond bars are a function of the tensile strength of concrete (f_{ctk}).

The guidance for detailing different types of members includes requirements for minimum areas of reinforcement. This is stipulated in order to (1) prevent a brittle failure, (2) prevent wide cracks and (3) resist stresses arising from temperature effects, shrinkage and other restrained actions.

10.2. Discussion of the general requirements

In this section the main features of the detailing requirements are arranged in a practical order and discussed.

10.2.1. Cover to bar reinforcement

See Chapter 9 of this guide for the minimum cover to reinforcement to meet durability requirements. See Chapter 12 for cover requirements for fire resistance.

10.2.2. Spacing of bars

The basic principle is that the reinforcement bars in a member should be arranged in such a way that concrete can be placed and compacted satisfactorily so that adequate bond will develop between the bars and concrete.

Figure 10.1 defines the spacing $S_{\min}$ between bars in a layer and that between layers of reinforcement. $S_{\min}$ should be taken as diameter of the bar or diameter of the aggregate plus 5 or 20 mm, whichever is the largest.

10.2.3. Mandrel diameters for bars

EN 1992-1-1 prescribes separate mandrel diameters for (1) avoiding damage to the reinforcement as a result of bending the bars and (2) for avoiding damage to the concrete when a hook or a bend carries stress.

For avoiding damage to the bars themselves, the diameter of the mandrel is 4ϕ for bars of 16 mm diameter or less and 7ϕ for larger bars.

Clause 8.3(3)

When a bent bar carries stress, compressive stresses are generated inside the bend. For a given force the stress inside the bend is inversely related to the diameter of the bend (i.e. the stress decreases with an increase in the diameter of the bend). EN 1992-1-1 gives a formula relating the diameter of the mandrel, diameter of the bar, the force in the bar and the spacing of the bars. Table 10.1 sets out the same information in a readily usable form for different bar spacing and concrete grades.

10.2.4. Basic anchorage length

In EN 1992-1-1 the parameter $l_{\text{b, reqd}}$ (basic anchorage length) is used in the calculation of design anchorage and lap lengths:

$$l_{\text{b, reqd}} = (\phi/4)(\sigma_{\text{sd}}/f_{\text{bd}})$$

where ϕ is the diameter of the bar, σ_{sd} is the stress in the bar and f_{bd} is the design value of ultimate bond stress. The axial force in the bar is assumed to be distributed uniformly over this length through bond stress between the reinforcement and concrete.

The ultimate bond stress basically depends upon the tensile strength of concrete and the location of the bar within the concrete. The latter is referred to as 'bond condition'. EN 1992-1-1 defines 'good' and 'poor' bond conditions, and these are shown in Fig. 10.2. There is also test evidence to show that the ultimate bond stress has some dependence on the size of the bar. EN 1992-1-1 de-rates the bond stress for large-diameter bars. In the UK National Annex it is defined as bars larger than 40 mm. The above factors are reflected in the expression for the ultimate stress:

$$f_{\text{bd}} = 2.25\eta_1\eta_2(f_{\text{ctk, 0.05}}/\gamma_{\text{c}})$$

where η_1 is 1.0 for 'good' bond conditions and 0.7 for 'poor' bond conditions and $\eta_2 = 1.0$ for bar $\phi \le 32$ mm and $(132 - \phi)/100$ for bar $\phi > 32$ mm. In the UK this expression should strictly be $(140 - \phi)/100$.

Table 10.2 shows the ultimate bond stress for different grades of concrete together with values of basic anchorage lengths assuming full design stress in the bars.

10.3. Anchorage of longitudinal bars

When reinforcement is designed to carry stress, it needs to be anchored into adjacent parts such that (1) the required stress will be able to develop and (2) the force in the bar is safely transmitted to the surrounding concrete without causing longitudinal cracks or spalling.

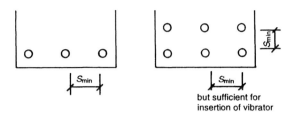

Fig. 10.1. Spacing of bars

Table 10.1. Mandrel diameters of curved bars as multiples of bar diameter

a_b	Concrete grade (MPa)							
	20	25	30	35	40	45	50	≥ 55
2ϕ	25.6	20.5	17.1	14.6	12.8	11.4	10.2	9.3
3ϕ	21.3	17.0	14.2	12.1	10.6	9.5	8.5	7.7
4ϕ	19.2	15.4	12.8	11.0	9.6	8.6	7.7	7.0
5ϕ	18.0	14.4	12.0	10.2	9.0	8.0	7.0	6.5
10ϕ	15.4	12.3	10.3	8.8	7.7	6.8	6.1	5.6

(1) a_b is half the centre-to-centre distance between bars. For a bar adjacent to the face of a member, a_b is the cover plus half the diameter of the bar.
(2) The table assumes that the stress in the bar is f_{yk}/γ_s, i.e. 500/1.15 MPa. The table values may be multiplied by $A_{s,\,reqd}/A_{s,\,provided}$.

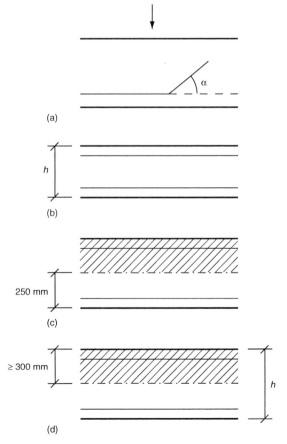

Fig. 10.2. Bond conditions: (a) $45° < \alpha < 90°$ for all h values; (b) $h < 250$ mm; (c) 250 mm $< h < 600$ mm; (d) $h > 600$ mm (hatched zones have poor bond conditions)

Generally, transverse reinforcement will be required in the anchorage zone to resist the secondary forces induced locally.

Normal methods of anchorage are shown in Fig. 10.3.

Table 10.2. Bond stress and basic anchorage length for different concrete grades

	Concrete grade (MPa)								
	20	25	30	35	40	45	50	55	≥ 60
f_{bd} (MPa) ('good' bond conditions)	2.25	2.70	3.00	3.30	3.75	4.05	4.35	4.50	4.65
$l_{b, rqd}$ ('good' bond conditions)	48	40	36	33	29	27	25	24	23
$l_{b, rqd}$ ('poor' bond conditions)	69	58	52	47	41	38	36	35	33

(1) For 'poor' bond conditions the table values should be multiplied by 0.7.

(2) The table values apply to bar diameters $\phi \leq 40$ mm. For larger bars the values should be multiplied by $(140 - \phi)/100$.

(3) The basic anchorage length $l_{b, rqd}$ has been calculated assuming that the stress in the bar is f_{yk}/γ_s, i.e. 500/1.15. The values of $l_{b, rqd}$ may be multiplied by $A_{s, reqd}/A_{s, provided}$.

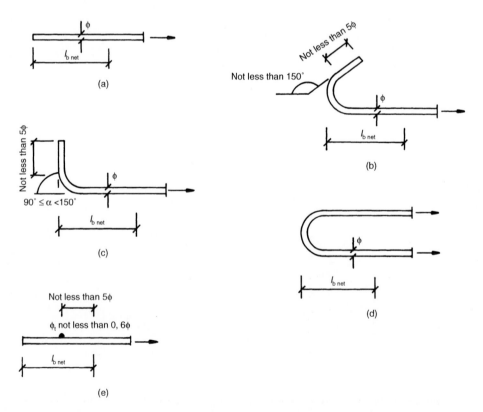

Fig. 10.3. Methods of anchorage: (a) straight bar; (b) hook; (c) bend; (d) loop; (e) welded transverse bar

10.3.1. Design anchorage length

The basic anchorage length can be modified to allow for such effects as the shape of the bars, the size of the concrete cover, confinement offered by transverse reinforcement or transverse pressure. In EN 1992-1-1, the design anchorage length is obtained by multiplying the basic anchorage length by a number of factors:

$$l_{bd} = \alpha_1\alpha_2\alpha_3\alpha_4\alpha_5 l_{b,reqd} > \max\{0.3l_{b,reqd}; 10\phi; 100 \text{ mm}\}$$

for anchorage in tension and

$$l_{bd} = \alpha_1\alpha_2\alpha_3\alpha_4\alpha_5 l_{b,reqd} > \max\{0.6l_{b,reqd}; 10\phi; 100 \text{ mm}\}$$

for anchorage in compression; the product $\alpha_2\alpha_3\alpha_5$ should be ≥ 0.7. Values of the different multipliers together with the conditions that should be met are given in the code and are not reproduced here. However, the applicability of these in practice are discussed below.

The following applies to bars in tension:

- α_1 allows for the shape of the bars. For straight bars, $\alpha_1 = 1.0$. For curved bars, α_1 may be taken as 0.7, provided that the lesser of the side cover to the bars or half the clear spacing between bars is $\geq 3\phi$. It will be difficult to meet this condition in all but a few practical cases.
- α_2 allows for the effect of the size of concrete cover. For straight bars in tension, some reduction will be possible when the parameter c_d is between ϕ and 3ϕ, where c_d is the least of the side or bottom cover or half the clear spacing between the bars. In the case of curved bars this benefit does not accrue until c_d is larger than 3ϕ. Again, advantage can be taken of this reduction only in a limited number of practical cases.
- α_3 allows for the effect of confinement offered by transverse reinforcement, which is not welded to main bars. The transverse reinforcement should be placed between the concrete surface and the bar that is being anchored. The reduction will be enhanced if the transverse reinforcement is in the form of links. Even so, the reduction that can be achieved in beams will only be 0.925 when $\sum A_{st} = 1.0A_s$.
- $\alpha_4 = 0.7$ can be used in all cases where the transverse reinforcement is welded to the main bars, provided the diameter of the transverse bar is at least 0.6ϕ and it is located at least 5ϕ inside $l_{b,reqd}$ from the free end of the bar.
- α_5 accounts for the effect of any pressure p (in megapascals) transverse to the potential plane of splitting, and is taken as $(1 - 0.04p)$.

In compression anchorage, $\alpha_1 = \alpha_2 = \alpha_3 = \alpha_4 = 1.0$; α_5 does not apply to anchorage in compression.

In summary, the conditions that need to be satisfied to take advantage of the reduction factors are such that they will only apply in a limited number of practical cases in building structures.

10.3.2. Transverse reinforcement at anchorage

In *clause 8.4.1*, EN 1992-1-1 states that transverse reinforcement should be provided if necessary. There is no other direct guidance. In the section dealing with design anchorage length there are references to transverse reinforcement. This could be interpreted to mean that there is no specific requirement for transverse reinforcement. This is unlike the requirements in the ENV, which still appear to be prudent, and are repeated here:

Clause 8.4.1

- In beams, transverse reinforcement should be provided for anchorage in tension, if there is no transverse compression due to support reaction (e.g. indirect supports); and for all compression anchorage.
- The minimum total area of the transverse reinforcement (the area of the legs parallel to the plane of the longitudinal reinforcement) is 25% of the area of one anchored bar. (Fig. 10.4).

- The transverse reinforcement should be distributed evenly along the anchorage length, and at least one bar should be placed in the region of the hook, bend or loop of any curved bars.
- For bars in compression, transverse reinforcement should surround the bars, being concentrated at the end of the anchorage and extend beyond it to a distance of at least four times the diameter of the anchored bar. This is a precaution against bursting forces that are likely to arise due to the 'pin effect' at the end of the bar.

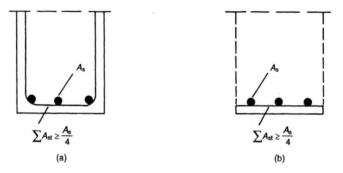

Fig. 10.4. Lateral reinforcement: (a) beam; (b) slab (A_{st} is the area of one bar of the transverse reinforcement; A_s is the area of one anchored bar)

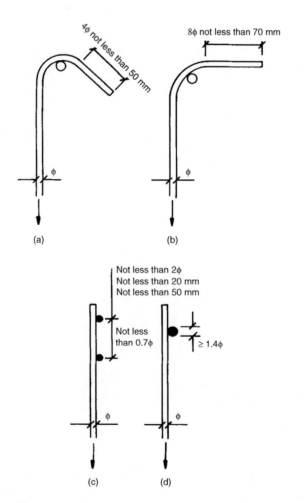

Fig. 10.5. Anchorage of links

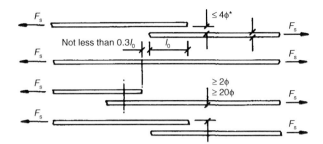

*If > 4φ, the lap length shall be increased by the amount by which the clear gap exceeds 4φ

Fig. 10.6. Spaces between lapped bars

10.4. Anchorage of links

These are shown in Fig. 10.5.

10.5. Laps for bars

10.5.1. General

As the length of reinforcement bars is generally likely to be less than the length of the structure, lapping (or splicing) of bars will be necessary in most structural elements. At laps, forces are transmitted from one bar to another. This can be achieved through the concrete surrounding the lapping bars or by welding of the bars or by mechanical couplers.

Lapped bars should comply with the arrangement shown in Fig. 10.6.

10.5.2. Design lap length

Design lap length is calculated by modifying the basic anchorage length (see Section 10.2.4 above) by applying multipliers:

$$l_0 = \alpha_1\alpha_2\alpha_3\alpha_5\alpha_6 l_{b, rqd} \geq \max\{0.3\alpha_6 l_{b, rqd}; 15\phi; 200 \text{ mm}\}$$

where α_1, α_2, α_3 and α_5 are as discussed for anchorage and α_6 allows for the effect of the amount of reinforcement that is lapped at one section:

$$\alpha_6 = (\rho_1/25)^{0.5}$$

with values in the range 1.0–1.5 (for $\rho_1 > 50\%$). The amount of reinforcement lapped at one section includes any laps occurring within $0.65l_0$ on either side of the section.

10.5.3. Transverse reinforcement

The requirements are shown in Fig. 10.7.

10.6. Additional requirements for large diameter bars

Clause 8.8 applies to bars with a diameter larger than 32 mm, but in the UK to bars > 40 mm. Such bars should be anchored either as straight bars with links provided as confining reinforcement or using mechanical devices.

Clause 8.8

Large-diameter bars should not be lapped unless the minimum dimension of the member is 1 m, or the stress in the bar does not exceed 80% of the design strength.

In the absence of transverse compression, transverse reinforcement should be provided in anchorage zones. This is in addition to any shear reinforcement provided. The area of transverse reinforcement is shown in Fig. 10.8.

Unless cracking is verified by calculations, surface reinforcement should be provided such that its area is $0.02A_{ct, ext}$ in the direction parallel to the large diameter bars and $0.01 A_{ct, ext}$ in the direction perpendicular to it. $A_{ct, ext}$ is the area of concrete between the links and the outer surface of the member below the neutral axis. Surface reinforcement is not practical and should be avoided.

10.7. Requirements for weld mesh

The mandrel sizes for bends in mesh are shown in Fig. 10.9.

The methods of lapping are shown in Fig. 10.10.

Intermeshed laps should comply with the requirements for bars discussed in Section 10.5 above. In the calculation of l_0, the value of α_3 should be taken as 1.0.

Layered laps of main bars in a mesh should be generally situated in zones, where the calculated stress in the main reinforcement does not exceed 80% of its design strength.

The permissible percentage of the main reinforcement that can be lapped in one section is as follows:

- for intermeshed laps of fabrics using high bond bars: 100%
- for layered laps with $A_s/S \leq 1200$ mm^2/m: 100%
- for layered laps with $A_s/S > 1200$ mm^2/m: 60%.

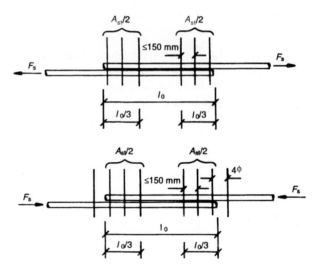

Fig. 10.7. Placing of transverse reinforcement

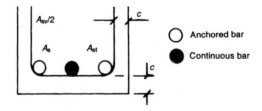

Fig. 10.8. Additional transverse reinforcement

$d \geq 3$: 5ϕ
$d < 3$ or welding within the curved zone: 20ϕ

5ϕ

Note: the mandrel size for welding within the curved zone may be reduced to 5ϕ where the welding is carried out in accordance with prEN ISO 17660, Annex B

Fig. 10.9. Minimum diameter of the mandrel: (a) welds outside bends; (b) welds inside bends

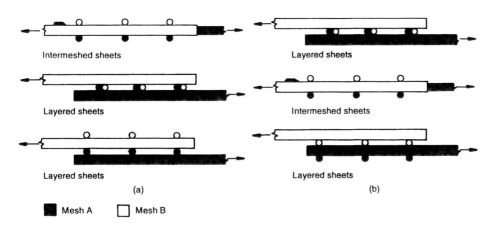

Fig. 10.10. Layering of wires: (a) main reinforcement; (b) transverse reinforcement

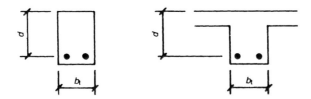

Fig. 10.11. Minimum area of longitudinal reinforcement

The permissible percentage of secondary reinforcement that can be lapped in one section is 100%.

10.8. Bundled bars

Unless stated otherwise, the rules given above for individual bars also apply to bundled bars.

The bundle is replaced by a notional bar of diameter $\phi_n = \phi\sqrt{n_b}$, in which n_b is the number of bars with diameter ϕ. n_b is limited to 4 or below for vertical bars in compression and for bars in a lapped joint; and 3 or below in all other cases.

In good bond conditions, two touching bars placed one over the other need not be considered as a bundle.

Individual bars in bundles with $\phi_n \geq 32$ mm should be anchored in tension such that they are staggered by $1.3l_{b, reqd}$ in the longitudinal direction. For the calculation of $l_{b, reqd}$, the diameter of the individual bar may be used.

In compression anchorages of bundles with $\phi_n \geq 32$ mm at least four links with bars of 12 mm diameter or more should be provided.

In lapped joints of bundles comprising two bars and $\phi_n \geq 32$ mm or of lapped joints in bundles comprising three bars, laps of the individual bars in the bundle should be staggered in the longitudinal direction by at least $1.3l_0$.

10.9. Detailing requirements for particular member types

10.9.1. Beams

Longitudinal reinforcement

(1) Minimum area $A_{st, min}$:
  - $A_{st, min} = 0.26 f_{ctm} b_t d/f_{yk}$ but not less than $0.0013 b_t d$, where f_{yk} is the characteristic yield stress of reinforcement (Fig. 10.11).

-   At supports in monolithic construction where simple supports are assumed in the design, reinforcement $A_{st,\,sup}$ required to cope with partial fixity is $0.15 \times A_{st}$ span (Figs 10.12 and 10.13).
(2) Maximum area $A_{st,\,max}$ or $A_{sc,\,max}$:
    -   $A_{st,\,max}$ or $A_{sc,\,max} = 0.04A_c$ where A_c is the cross-section area of concrete.
(3) Distribution of support reinforcement at internal supports of continuous flanged beams. Total support reinforcement (A_{st}) should be distributed over the width of the effective flange. Part of it may be concentrated over the width of the web (Fig. 10.14).
(4) Longitudinal compression reinforcement (diameter ϕ) should be contained by link reinforcement, the maximum spacing of which should not exceed 15ϕ.

Shear reinforcement
(1) General:
    -   Shear reinforcement should form an angle of 90–45° with the mid-plane of the beam.
    -   Shear reinforcement may consist of a combination of:
        -   links enclosing the longitudinal tensile reinforcement and the compression zone
        -   bent-up bars
        -   shear assemblies of cages, ladders, etc. which do not enclose the longitudinal reinforcement, but which are properly anchored in the compression and tension zones (Fig. 10.15).
    -   For combinations of links and shear assemblies, all shear reinforcement should be effectively anchored. Lap joints on the leg near the surface of the web are only permitted for high-bond bars. At least 50% of the necessary shear reinforcement should be in the form of links.

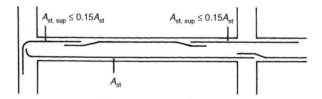

Fig. 10.12. Longitudinal reinforcement at supports in monolithic construction

Fig. 10.13. Maximum area of longitudinal reinforcement

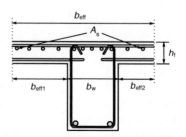

Fig. 10.14. Placing of tension reinforcement in flanged cross-section

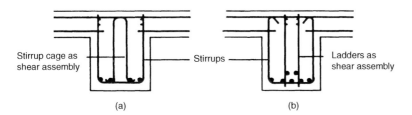

Fig. 10.15. Combination of links and shear assemblies: (a) stirrup cage as a shear assembly; (b) ladders as a shear assembly

(2) Minimum area A_{sw}:
- The minimum area should be calculated from

$$\rho_{w, min} = A_{sw}/sb_w \sin \alpha = 0.08f_{ck}^{0.5}/f_{yk}$$

where $\rho_{w, min}$ is the minimum shear reinforcement ratio, A_{sw} is the area of shear reinforcement with a longitudinal spacing s and α is the angle between the shear reinforcement and the longitudinal steel.

(3) Maximum spacing of shear reinforcement S_{max}:
- Maximum longitudinal spacing of links $= 0.75d(1 + \cot \alpha)$.
- Maximum longitudinal spacing of bent-up bars $= 0.6d(1 + \cot \alpha)$.

(4) The transverse spacing of a series of shear links should not exceed $0.75d$ nor 600 mm.

Curtailment of longitudinal reinforcement

Any curtailed reinforcement should be provided with an anchorage length $l_{b, net}$ but not less than d from the point where it is no longer needed. This should be determined taking into account the tension caused by the bending moment and that implied in the truss analogy used for shear design. This can be done by shifting the point of the theoretical cut-off based on the bending moment by a_1 (see below for definition) in the direction of decreasing moment. This procedure is also referred to as the 'shift rule'.

$$a_1 = z(\cot \theta - \cot \alpha)/2$$

where θ is the angle of the concrete struts to the longitudinal axis and α is the angle of the shear reinforcement to the longitudinal axis. Normally, z can be taken as $0.9d$. For reinforcement in the flange, placed outside the web, a_1 should be increased by the distance of the bar from the web.

Anchorage at supports

(1) End support:
- When there is little or no fixity at an end support, at least one-quarter of the span reinforcement should be carried through to the support. The code recommends that the bottom reinforcement should be anchored to resist force of $V_{sd}a_1/d + N_{sd}$, where V_{sd} is the shear force at the end, a_1 is as defined above for the shift rule, and N_{sd} is the axial force, if any, in the member.

(2) Intermediate supports (general requirements):
- At intermediate supports, $\geq 25\%$ of the midspan bottom reinforcement should be carried to the support.
- The minimum anchorage of bottom reinforcement beyond the face of the support is 10ϕ for straight bars; or the diameter of the mandrel for bars of 16 mm diameter or more and with hooks or bends; or twice the diameter of the mandrel in other cases.
- However, this does not mean that the support must be greater than 20ϕ wide, as the bars from each side of the support can be lapped. It is, however, recommended that continuous reinforcement is provided to resist accidental forces (Fig. 10.16).

Skin reinforcement

Skin reinforcement to control cracking should normally be provided in beams over 1.0 m in depth where the reinforcement is concentrated in a small portion of the depth. This reinforcement should be evenly distributed between the level of the tension steel and the neutral axis and be located within the links.

Surface reinforcement

Surface reinforcement may be required to resist spalling of the cover, e.g. arising from fire or where bundled bars or bars greater than 32 mm are used. This reinforcement should consist of small-diameter, high-bond bars or wire mesh placed in the tension zone outside the links.

The area of surface reinforcement parallel to the beam tension reinforcement should not be less than $0.01A_{ct, ext}$, where $A_{ct, ext}$ is the area of concrete in tension external to the links.

The longitudinal bars of the surface reinforcement may be taken into account as longitudinal bending reinforcement, and the transverse bars as shear reinforcement, provided that they meet the arrangement and anchorage requirements of these types of reinforcement (Fig. 10.17).

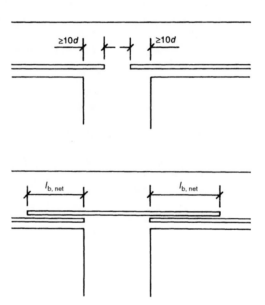

Fig. 10.16. Continuous reinforcement

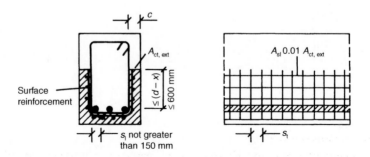

Fig. 10.17. Arrangement and anchorage requirements

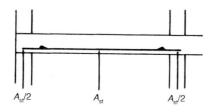

Fig. 10.18. Span reinforcement

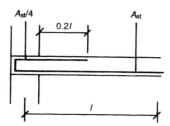

Fig. 10.19. End supports with partial fixity

10.9.2. Slabs
Longitudinal reinforcement
(1) Minimum area $A_{st, min}$:

$$A_{st, min} = 0.26 f_{ctm} b_t d / f_{yk}$$

but not less than $0.0013 b_t d$, where f_{yk} is the characteristic yield stress of reinforcement.
(2) Maximum area $A_{st, max}$:

$$A_{st, max} \geq 0.04 A_c$$

where A_c is the cross-section area of concrete.
(3) Maximum spacing S_{max}:
  - Generally, $S_{max} = 3h \leq 400$ mm for main reinforcement and $3.5h \leq 450$ mm for secondary reinforcement.
  - Local to concentrated loads, $S_{max} = 2h \leq 250$ mm for main reinforcement and $3h \leq 400$ mm for secondary reinforcement.
(4) Reinforcement near supports:
  - In simply supported slabs a minimum of 50% of the reinforcement in the span should be anchored at supports (Fig. 10.18).
  - Where partial fixity is likely to exist despite the assumption of simple support in design, 25% of the reinforcement required to resist the maximum span moment should be provided at the top of end supports.
  - At the end supports, the reinforcement should extend from the face of the support, at least 0.2 times the adjacent span (Fig. 10.19).
  - At intermediate supports, the reinforcement should be continuous across the support.

Transverse reinforcement
The minimum area A_s is 20% of the longitudinal reinforcement.

Corner and edge reinforcement
Suitable reinforcement is required where slab corners are restrained against lifting. Normally, U bars extending $0.2l$ into the span should be provided at all edges (Fig. 10.20).

Shear reinforcement

(1) Minimum slab depth. $h_{min} = 200$ mm where shear reinforcement is to be provided.
(2) General:
  – The requirements given in Section 10.9.1 for beams apply generally to slabs, with the following modifications:
    – Form of shear reinforcement: shear reinforcement may consist entirely of bent-up bars or shear assemblies where $V_{Ed} \leq 0.33 V_{Rd\,max}$.
    – Maximum longitudinal spacing:

$$s_{max} = 0.75d(1 + \cot \alpha)$$

    for links and $1.0d$ for bent up bars.
    – Maximum transverse spacing of shear reinforcement is $1.5d$.

10.9.3. Columns

Longitudinal reinforcement

(1) Minimum diameter: 12 mm.
(2) Minimum area $A_{s,\,min}$:

$$A_{s,\,min} = 0.1 N_{Ed}/f_{yd} \quad \text{or} \quad 0.002 A_c$$

whichever is greater, where N_{Ed} is the design axial force, f_{yd} is the yield strength of reinforcement and A_c is the area of concrete.
(3) Maximum area $A_{s,\,max}$:

$$A_{s,\,max} = 0.04 A_c$$

outside laps and

$$A_{s,\,max} = 0.08 A_c$$

at laps.
(4) Minimum number of bars (Fig. 10.21).

Transverse reinforcement

(1) General:
  – All transverse reinforcement must be adequately anchored.
  – Every longitudinal bar (or group of bars) placed in a corner should be held by transverse reinforcement.
  – No longitudinal bar in a compression zone should be further from a restrained bar than 150 mm.
(2) Minimum diameter:
  – The diameter of the transverse bar should not be less than 6 mm or 0.25 times the diameter of the largest bar being restrained.
(3) Spacing:
  – Generally, the maximum spacing S_{max} should be the least of the following: 20 times the diameter of the longitudinal bar; or the lesser dimension of the column; or 400 mm.

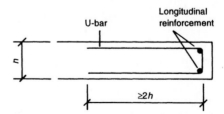

Fig. 10.20. Reinforcement at free edges

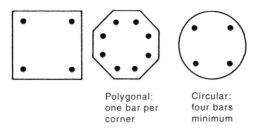

Polygonal: Circular:
one bar per   four bars
corner   minimum

Fig. 10.21. Minimum number of bars

- For a distance equal to the larger dimension of the column, above and below slabs or beams the spacing noted above should be reduced by a factor of 0.6.
- The above reduced spacing is also required at laps of longitudinal bars of diameter greater than 14 mm. A minimum of three transverse bars should be evenly positioned over the lap length.

10.9.4. Walls

Vertical reinforcement

(1) Minimum area $A_{sv, min}$

$$A_{sv, min} = 0.002A_c$$

(2) Maximum area $A_{sv, max}$:

$$A_{sv, max} = 0.04A_c$$

The code permits this to be doubled if the designer can show that the integrity of concrete is not affected and the full strength can be achieved at the ultimate limit state.

(3) Maximum spacing S_{max}:
- The distance between adjacent bars should not exceed three times the thickness of the wall or 400 mm, whichever is less.

(4) When minimum reinforcement controls the design, 50% of the minimum reinforcement should be placed on each face.

Horizontal reinforcement

(1) Placement:
- Horizontal reinforcement should be placed between the vertical reinforcement and the face of the wall.

(2) Minimum area $A_{sh, min}$:

$$A_{sh, min} = 0.25A_{sv}$$

but not less than $0.001A_c$.

(3) Maximum spacing S_{max}:

$$S_{max} = 400 \text{ mm}$$

(4) Minimum diameter:
- The code does not specify a value, but it will be prudent to use a minimum of 0.25 times the diameter of the vertical reinforcement.

Transverse reinforcement

Where the area of vertical reinforcement exceeds $0.02A_c$ transverse reinforcement in the form of links should be provided in accordance with the requirements for columns.

10.9.5. Corbels

General
Where $a_c \leq z_0$, a simple strut-and-tie model may be used (Fig. 10.22a).

Anchorage of the primary tie reinforcement
Unless a length l_{bnet} is available, the primary horizontal tie A_s should be anchored on both sides beyond the bearing area using U bars or a welded cross bar.

Provision of links
When $a_c < 0.5h_c$, closed horizontal or inclined links should be provided (Fig. 10.22b). The area of the link should be at least 0.25 times the area of the primary tie reinforcement.

When $a_c \geq 0.5h_c$ and the load applied on the corbel exceeds V_{Rdc}, vertical closed links should be provided in addition to the horizontal links (Fig. 10.22c). The area of the vertical link should be at least $0.5F_{wd}/f_{yd}$, where F_{wd} is the force in the main tie reinforcement.

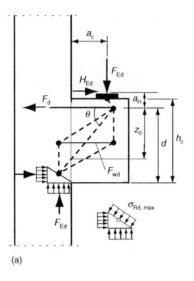

(a)

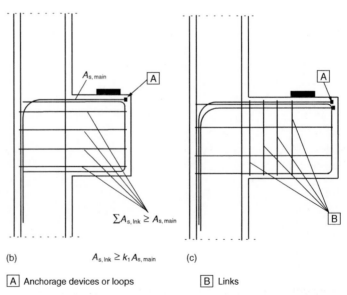

(b) $A_{s,\,lnk} \geq k_1 A_{s,\,main}$ (c)

| A | Anchorage devices or loops | B | Links |

Fig. 10.22. Corbels: (a) strut-and-tie model; (b) reinforcement for $a_c < 0.5h_c$; (c) reinforcement for $a_c \geq 0.5h_c$

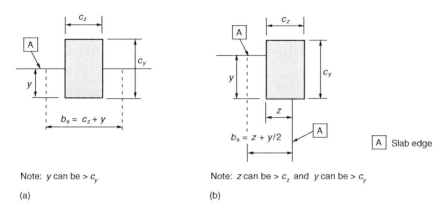

Fig. 10.23. Effective width, b_e, of a flat slab: (a) edge column; (b) corner column

10.9.6. Nibs

The code does not give any specific guidance for designing or detailing such members, which normally project from the faces of beams or walls. Nibs less than 300 mm thick may be designed as short cantilevers. The line of action of load should be taken as the outer edge of the loaded area. The bending moment should be calculated on the line of the nearest vertical reinforcement in the member from which the nib projects (i.e. the leg of a link in a beam or vertical reinforcement in a wall). Nibs should be sized so that shear reinforcement is not required. Reduction of flexural shear permitted in *clause 6.2.2(6)* could be taken into account in assessing the reinforcement. The tension reinforcement should be adequately anchored by forming loops in the vertical or horizontal plane. Where nibs hang from the bottom of other members, sufficient tension reinforcement should be provided to transfer the loads from the nib to the member.

Clause 6.2.2(6)

10.9.7. Reinforcement in flat slabs

Concentration of reinforcement over the columns will generally be required to meet the serviceability requirements. In the absence of rigorous calculations, top reinforcement with an area of $0.5A_t$, should be placed in a width equal to the sum of 0.125 times the panel width on either side of the column. A_t represents the area of reinforcement required to resist the full negative moments in a panel.

At least two bars forming the bottom reinforcement in the slab should pass through the internal columns in each orthogonal direction.

Reinforcement required to transfer bending moments from slab to columns, at right angles to an edge, should be placed within an effective width as shown in Fig. 10.23.

EN 1992-1-1 recognizes the use of both proprietary shear reinforcement and conventional link reinforcement. In the case of the former, the design and detailing should comply with the relevant European technical approval.

Where punching shear reinforcement is required, it should be provided between the loaded area and $1.5d$ within the control perimeter at which shear reinforcement is no longer required. Such reinforcement should be provided in at least two perimeters of spacing not exceeding $0.75d$. The spacing of the perimeters in which links are provided should not exceed $0.75d$. The spacing of the legs of links around a perimeter should not exceed $1.5d$ (Fig. 10.24).

Where shear reinforcement is required, the area of one leg of link reinforcement should comply with

$$A_{sw, min}(1.5 \sin \alpha + \cos \alpha)/s_r s_t \geq 0.08 f_{ck}^{0.5}/f_{yk}$$

where α is the angle between the shear reinforcement and the longitudinal steel, and s_r and s_t are the spacings of the shear reinforcement in the radial and tangential directions, respectively.

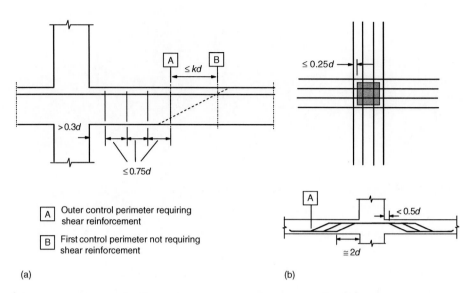

Fig. 10.24. Punching shear reinforcement: (a) spacing of links; (b) spacing of bent-up bars

Bent-up bars passing through the loaded area and within $0.25d$ on either side of it may be used as punching shear reinforcement.

CHAPTER 11

Prestressed concrete

11.1. General

In EN 1992-1-1, prestressed concrete is treated as part of a much wider material group known as reinforced concrete, which covers normal reinforced concrete through partially prestressed concrete to fully prestressed concrete. The principles and methods given in EN 1992-1-1 are, in general, applicable to the full range of reinforced concrete construction. This chapter covers the aspects of design and detailing that are particular to prestressed concrete; those that apply equally to reinforced concrete are covered in other chapters of this guide.

EN 1992-1-1 covers the design of prestressed concrete members in which the prestress is applied by tendons; the tendons may be embedded in the concrete and may be pretensioned and bonded or post-tensioned and bonded or unbonded. The tendons may also be external to the structure with points of contact occurring at deviators and anchorages. It should be noted that second-order moments can be induced in prestressed members with external tendons if their points of contact with the concrete are sufficiently far apart.

Clause 2.3.1.4

11.2. Summary of main clauses

The main clauses relating to the design of prestressed concrete in EN 1992-1-1 are summarized in Table 11.1.

11.3. Durability

Steel reinforcement and prestressing tendons are protected against corrosion by complying with EN 1992-1-1 requirements on stress levels, crack widths and concrete cover. EN 1992-1-1 specifies a 'minimum' cover, based on the greater of the requirements for bond or durability, to which a construction tolerance of 10 mm must be added.

Clause 4.4.1.3(1)P

For post-tensioned members, the minimum cover for bond requirements should not be less than the duct diameter (for circular ducts), or the greater of the smaller dimension or half the larger dimension (for rectangular ducts). In no case does it need to be greater than 80 mm. Although EN 1992-1-1 specifies requirements for the minimum cover perpendicular to the plane of curvature for curved reinforcing bars, it does not specify similar requirements for curved prestressing tendons. This is an unfortunate omission, and the designer is recommended to refer to other codes, e.g. BS 8110, where specific requirements are stated for the prevention of bursting of the cover perpendicular to the plane of curvature. The requirements of BS 8110 are given in Table 11.2.

Clause 4.4.1.2(3)

For pretensioned members, the minimum cover for bond requirements should be not less than 1.5 the tendon diameter, or 2.5 times the diameter when indented wires are used. The

Clause 4.4.1.2(3)

minimum cover is not related to the aggregate size, but this will probably not be a problem in practice when the construction tolerance is taken into account.

The minimum cover to prestressing tendons and concrete strengths for durability requirements are given in *Tables 4.3N, 4.5N* and *E.1N*. The UK National Annex replaces these tables with two tables specifying minimum covers and concrete strengths, maximum water/cement ratios, minimum cement content, cement/combination types and equivalent designated concrete for different exposure classes and for intended working lives of 50 and

Table 11.1. Clauses in EN 1992-1-1 for the design of prestressed concrete members

Requirement	Clause/annex
Durability	
Environmental conditions	4.2
Concrete cover	4.4.1
Design data	
Concrete	3.1
Prestressing steel	3.3
Mechanical properties	3.3.2–3.3.6
Jacking force	5.10.2, 5.10.3
Partial factors for prestress	2.4.2.2, 5.10.9
Ultimate limit state	
Bending and longitudinal force	6.1
Values of prestress	2.4.2.2, 5.10.3, 5.10.8
Shear	
General	6.2.1
Reduction in web width	6.2.3(6)
Shear capacity of concrete	6.2.2
Maximum design shear stress	6.2.1(6), 6.2.2(6), 6.2.3(3), 6.2.3(4)
Shear reinforcement	6.2.3
Enhancement near supports	6.2.2(6), 6.2.3(8)
Variable depth members	6.2.1(2)
Inclined tendons	6.2.1(3)
Torsion	6.3
Serviceability limit state	
Values of prestress	5.10.3, 5.10.9
Stress levels	7.1, 7.2
Cracking	7.3
Deformation	7.4
Prestress losses	
Relaxation	3.3.2, 5.10.6, 10.5.2, Annex D
Elastic deformation	5.10.4, 5.10.5.1
Shrinkage	3.1.4(6), 5.10.6, Annex B
Creep	3.1.4(2)–3.1.4(5), 5.10.6, Annex B
Draw-in	5.10.4, 5.10.5.3
Duct friction	5.10.4, 5.10.5.2
Anchorage zones	
Pretensioned members	6.5, 8.10.2
Post-tensioned members	6.5, 6.7, 8.10.3
Detailing	
Spacing of tendons/ducts	8.10.1
Anchorages and couplers	8.10.2–8.10.4
Deviators	8.10.5
Minimum area of tendons	5.10.1(6), 7.3.2, 9.2.1.1, 9.3.1.1
Tendon profile	9.2.1.3
Minimun shear reinforcement	9.2.2(5), 9.3.2
Spacing of shear reinforcement	9.2.2(6)–9.2.2(8), 9.3.2

Table 11.2. Minimum cover to curved ducts perpendicular to the plane of curvature

Minimum cover (mm):

Radius of curvature of duct (m)	19	30	40	50	60	70	80	90	100	110	120	130	140	150	160	170
Duct internal diameter (mm): Tendon force (kN):	296	387	960	1337	1920	2640	3360	4320	5183	6019	7200	8640	9424	10338	11248	13200
2	50	55	155	220	320	445										
4		50	70	100	145	205	265	350	420					Radii not normally used		
6			50	65	90	125	165	220	265	310	375	460				
8				55	75	95	115	150	185	220	270	330	360	395		
10				50	65	85	100	120	140	165	205	250	275	300	330	
12					60	75	90	110	125	145	165	200	215	240	260	315
14					55	70	85	100	115	130	150	170	185	200	215	260
16					55	65	80	95	110	125	140	160	175	190	205	225
18					50	65	75	90	105	115	135	150	165	180	190	215
20						60	70	85	100	110	125	145	155	170	180	205
22						55	70	80	95	105	120	140	150	160	175	195
24						55	65	80	90	100	115	130	145	155	165	185
26						50	65	75	85	100	110	125	135	150	160	180
28							60	75	85	95	105	120	130	145	155	170
30							60	70	80	90	105	120	130	140	150	165
32							55	70	80	90	100	115	125	135	145	160
34							55	65	75	85	100	110	120	130	140	155
36							55	65	75	85	95	110	115	125	140	150
38							50	60	70	80	90	105	115	125	135	150
40	50	50	50	50	50	50	50	60	70	80	90	100	110	120	130	145

(1) The tendon force shown is the maximum normally available for the given size of duct (taken as 80% of the characteristic strength of the tendon).

(2) Where tendon profilers or spacers are provided in the ducts and these are of a type which will concentrate the radial force, the values given in the table will need to be increased.

(3) The cover for a given combination of duct internal diameter and radius of curvature shown in the table may be reduced in proportion to the square root of the tendon force when this is less than the value tabulated, subject to a minimum value of 50 mm.

100 years (see Chapter 9). The tables are limited to exposure classes XC, XD and XS for a working life of 50 years and exposure class XC for a working life of 100 years because there is insufficient data to make recommendations for members exposed to other conditions. For selecting the quality of concrete subject to freeze–thaw and concrete in aggressive ground conditions, the designer is referred to Annex A of BS 8500-1.

The UK tables allow a greater range of concrete strengths and covers and take advantage of the improved durability of blended cements to reduce minimum concrete strengths and increase water/cement ratios for the same cover. As far as the minimum cover to ensure durability is concerned, the UK Annex does not distinguish between normal reinforcement and prestressing tendons. EN 1992-1-1, on the other hand, suggests that greater covers should be provided to prestressing steel. Presumably this is because it is generally believed that the risks posed to prestressing tendons from the breakdown of the protection provided by the concrete are greater. A simplified composite version of the tables in EN 1992-1-1 is presented as Table 11.3.

11.4. Design data

11.4.1. Concrete

Clause 3.1.3
Clause 3.1.2

EN 1992-1-1 does not specify minimum concrete strengths for prestressed concrete members. The design data to be assumed for concrete are the same for prestressed members as for normal reinforced members, and are given in Table 11.4.

11.4.2. Prestressing steel

Clause 3.3.2

Prestressing steel should comply with EN 10138.

Clause 3.3.6

EN 1992-1-1 adopts a bilinear stress–strain diagram for prestressing steel with either a horizontal or sloping top leg, as shown in Fig. 11.1.

Clause 5.10.2.1(1)

The maximum initial stress applied to the tendon, i.e. the jacking stress, is expressed as a percentage of the ultimate strength f_{pk} or the 0.1% proof stress $f_{p0.1k}$ of the tendon. In

Table 11.3. Minimum cover for durability requirements of prestressing tendons according to EN 1992-1-1

Exposure class	Working life (years)	Minimum cover (mm)						
		Minimum concrete strength:						
		C12/15	C20/25	C25/30	C30/37	C35/45	C40/50	C45/55
X0	50	10	10	10	10	10	10	10
	100	20	20	20	15	15	15	15
XC1	50	–	25	25	20	20	20	20
	100	–	35	35	30	30	30	30
XC2	50	–	–	35	35	30	30	30
	100	–	–	45	45	40	40	40
XC3	50	–	–	–	35	30	30	30
	100	–	–	–	45	40	40	40
XC4	50	–	–	–	40	40	35	35
	100	–	–	–	50	50	45	45
XD1, XS1	50	–	–	–	45	45	40	40
	100	–	–	–	55	55	50	50
XD2	50	–	–	–	50	50	45	45
	100	–	–	–	60	60	55	55
XS2	50	–	–	–	–	50	50	45
	100	–	–	–	–	60	60	55
XD3, XS3	50	–	–	–	–	55	55	50
	100	–	–	–	–	65	65	60

For values to be used in the UK, see Chapter 9 the National Annex.

Table 11.4. Concrete design data

Concrete class	f_{ck} (N/mm²)	$f_{ck, cube}$ (N/mm²)	f_{ctm} (N/mm²)	$f_{ctk0.05}$ (N/mm²)	$f_{ctk0.95}$ (N/mm²)	E_{cm} (kN/mm²)
C12/15	12	15	1.6	1.1	2.0	27
C16/20	16	20	1.9	1.3	2.5	29
C20/25	20	25	2.2	1.5	2.9	30
C25/30	25	30	2.6	1.8	3.3	31
C30/37	30	37	2.9	2.0	3.8	33
C35/45	35	45	3.2	2.2	4.2	34
C40/50	40	50	3.5	2.5	4.6	35
C45/55	45	55	3.8	2.7	4.9	36
C50/60	50	60	4.1	2.9	5.3	37

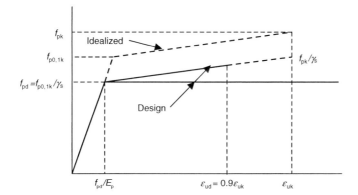

Fig. 11.1. Idealized and design stress–strain diagram for prestressing steel

Table 11.5. Jacking forces and initial prestress for different tendons*

Strand type	Diameter (mm)	Area (mm²)	f_{pk} (N/mm²)	$f_{p0.1k}$ (N/mm²)	Jacking force (kN)	Initial prestress (kN)
Y1770S7	15.2	140.0	1770	1520	192	181
	16.0	150.0	1770	1520	205	194
Y1860S7	12.5	93.0	1860	1600	134	126
	13.0	100.0	1860	1600	144	136
	15.2	140.0	1860	1600	202	190
	16.0	150.0	1860	1600	216	204
Y1860S7G	12.7	112.0	1860	1610	162	153
Y1820S7G	15.2	165.0	1820	1560	232	219
Y1700S7G	18.0	223.0	1700	1470	294	278

*Based on data from BS EN 10138-3

practice, $f_{p0.1k} \approx 0.85 f_{pk}$, and the requirement expressed in terms of the 0.1% proof stress will be the more critical. Similar requirements apply to the stress in the tendon after transfer. Values of jacking loads for various strand types are given in Table 11.5, where the requirement based on the 0.1% proof stress governs. *Clause 5.10.3(2)*

EN 1992-1-1 allows overstressing of tendons during tensioning to a stress of $0.95 f_{p0.1k}$, provided that the force in the jack can be measured to an accuracy of ±5% of the final value of the prestressing force. *Clause 5.10.2.1(2)*

Clause 2.4.2.2
Clause 5.10.9

11.4.3. Partial factors

The partial factors to be applied to the prestressing force in the design of prestressed concrete members are summarized in Table 11.6.

11.5. Design of sections for flexure and axial load

11.5.1. Ultimate limit state

Clause 2.4.2.2(1)
Clause 2.4.2.2(2)

EN 1992-1-1 specifies two partial safety factors for prestress when considering overall analysis and the design of sections: $\gamma_{p,\,fav} = 1.0$ and $\gamma_{p,\,unfav} = 1.3$. However, the UK National Annex specifies modified values of $\gamma_{p,\,fav} = 0.9$ and $\gamma_{p,\,unfav} = 1.1$. In most cases, prestressing is a favourable effect, and $\gamma_{p,\,fav}$ should be used.

In the design of sections, the initial strain in the tendons, ε_{p0}, multiplied by the partial factor, γ_p, is taken into account together with the increase in strain at the ultimate limit state, $\Delta\varepsilon_p$. The increase in strain at the ultimate limit state is determined as follows:

- For bonded tendons, the increase in strain in the tendon is equal to the increase in strain in the concrete above zero at the level of the tendon (Fig. 11.2).

Table 11.6. Partial factors to be used for prestress

		Partial factors			
		Serviceability limit state		Ultimate limit state	
Effect		Maximum	Minimum	Maximum	Minimum
Flexure					
Pretensioned or unbonded	EN 1992-1-1	1.05	0.95	1.3	1.0
	UK	1.0	1.0	1.1	0.9
Post-tensioned and bonded	EN 1992-1-1	1.1	0.9	1.3	1.0
	UK	1.0	1.0	1.1	0.9
Shear	EN 1992-1-1	–	–	1.3	1.0
	UK	–	–	1.1	0.9
Anchorage zones					
Anchorage of pretensioned tendons				See Section 11.8.1	
Anchorage zones				1.2	–

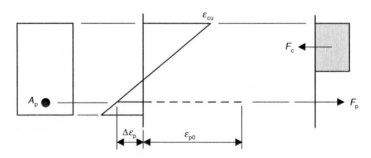

ε_{p0} = the initial strain in the tendon at a section (i.e. allowing only for losses due to friction and draw-in) + $\Delta\varepsilon_{p0}$

$\Delta\varepsilon_{p0}$ = 0, for pretensioned members
= the strain in the concrete due to stressing the tendon (or to stressing the first tendon when a number of tendons are stressed successively), for post-tensioned members

F_p = the force in the prestressing tendons at the ultimate limit state
= $E_p A_p (\Delta\varepsilon_p + \gamma_{p,\,fav} \varepsilon_{p0})$

Fig. 11.2. Strain in bonded prestressing tendons at the ultimate limit state

Table 11.7. Criteria for controlling cracking

Exposure class	Unbonded tendons: quasi-permanent loads	Bonded tendons	
		Frequent loads	Quasi-permanent loads
X0, XC1	0.3*	0.2	–
XC2, XC3, XC4	0.3	0.2	Decompression
XD1, XD2, XD3, XS1, XS2, XS3	0.3	Decompression	–

*For X0, XC1 exposure classes, crack width has no influence on durability and this limit state is set to guarantee acceptable appearance. In the absence of appearance conditions this limit may be relaxed.

- For unbonded and external tendons, the increase in strain has to be calculated based on the deformation of the concrete member. Alternatively, an increase in stress of 100 N/mm² may be assumed. When the deformation is calculated using mean values of the member properties, the increase in stress should be multiplied by a partial factor, $\gamma_{\Delta p}$. $\gamma_{\Delta p}$ should be taken as 0.8 or 1.2, depending on whether the increase in stress is a beneficial or adverse effect. Alternatively, a linear analysis using uncracked section properties can be used with $\gamma_{\Delta p} = 1.0$. *Clause 5.10.8(2)* *Clause 5.10.8(3)*

11.5.2. Serviceability limit state

EN 1992-1-1 distinguishes between three serviceability load combinations, i.e. characteristic, frequent and quasi-permanent. All are relevant to the design of prestressed concrete members, but their relative magnitude depends on the loads applied and the type of structure being designed.

EN 1992-1-1 also adopts an upper or lower characteristic value (r_{sup} and r_{inf}) for the prestressing force (whichever is more critical) at the serviceability limit state. The upper and lower characteristic values are taken as 1.05 and 0.95 times the mean value of the prestress respectively for pretensioned or unbonded tendons, and as 1.1 and 0.9 for post-tensioned bonded tendons. However, the UK National Annex adopts a single value of 1.0, which simplifies the analysis. In the design equations presented later in this section, r_{sup} and r_{inf} are explicitly identified so that the equations can be easily applied in other countries. *Clause 5.10.9(1)*

Limitation on stress

To avoid the occurrence of longitudinal cracks in regions of high compression, EN 1992-1-1 limits the compressive stress under the characteristic load combination to $0.6f_{ck}$ in areas exposed to environments of exposure class XD, XF or XS unless other measures, such as an increase in cover to reinforcement in the compressive zone or confinement by transverse reinforcement, are taken. In order to ensure that creep deformation remains linear, the concrete stress under the quasi-permanent load combination is limited to $0.45f_{ck}$. Otherwise, non-linear creep should be considered (see *clause 3.1.4*). Compressive stresses at transfer should generally be limited to $0.6f_{ck}$. However, for pretensioned members, this limit can be increased to $0.7f_{ck}$ if tests or experience demonstrate that longitudinal cracking is prevented. *Clause 7.2(2)* *Clause 7.2(3)* *Clause 3.1.4(4)* *Clause 5.10.2.2(5)*

To avoid stresses in the tendons under serviceability conditions that could lead to unacceptable cracking and deformation, EN 1992-1-1 limits the mean stress in the prestressing tendons to $0.75f_{pk}$, after allowance for losses. This is unlikely to be a critical criterion in most cases. *Clause 7.2(5)*

In calculating stresses at the serviceability limit state, sections should be assumed to be uncracked if the maximum tensile stress does not exceed the mean concrete tensile strength, f_{ctm}. *Clause 7.1(2)*

Cracking

EN 1992-1-1 specifies two criteria for the control of cracking: decompression and limiting crack widths to a specified value (Table 11.7). The decompression limit requires all parts of the tendon or duct to lie at least 25 mm within concrete in compression. Members *Clause 7.3.1(5)* *Clause 7.3.1(6)* prestressed with a combination of bonded and unbonded tendons should meet the requirements for bonded tendons. Cracking checks for prestressed concrete are carried out under the frequent or quasi-permanent load combinations using the upper or lower characteristic prestressing force.

In order to control cracking, EN 1992-1-1 tabulates bar size and spacing criteria (see *Tables 7.2N* and *7.3N*) to ensure that the crack widths are limited to the appropriate values, as long as a minimum area of reinforcement or prestressing tendons is provided. Alternatively, formulae are provided to allow the design crack width to be calculated.

Clause 7.3.3
Clause 7.3.3(2)
The approach adopted by EN 1992-1-1 for the design of partially prestressed concrete may be more rigorous than the approach adopted by some other codes (e.g. BS 8110), but it is more complex and difficult for the designer to apply. *Tables 7.2N* and *7.3N* can be applied to pretensioned members, where the crack width is controlled by the tendons, by using a steel stress equal to the increase in stress in the tendons due to the applied load, $\Delta\sigma_p$. They can also be used for post-tensioned members where the crack width is controlled by ordinary reinforcement by allowing for the effect of the prestress force when calculating the stress in the normal reinforcement.

Clause 7.3.4
Alternatively, the designer must use the crack width formulae. The procedure is as follows:

(1) Calculate the stress in the reinforcement, σ_s, assuming a cracked section and the design value of the prestress force $(r_{inf}P_{m\infty})$. For pretensioned members, σ_s can be replaced by the increase in tendon stress, $\Delta\sigma_p$.
(2) Determine $\varepsilon_{sm} - \varepsilon_{cm}$, allowing for tension stiffening using *equation (7.9)*.
(3) Determine $s_{r,\,max}$ using *equation (7.11)* or *(7.14)*.
(4) Calculate the design crack width w_k using *equation (7.8)*.
(5) Compare w_k with the required value, and adjust the normal reinforcement or prestressing force and/or its eccentricity.
(6) Repeat steps (1)–(5) until the required value of w_k is obtained.

Design equations
In order to assist the designer, the stress and crack width criteria discussed in the previous sections can be expressed mathematically as a series of inequalities. The following nomenclature is used (sagging moments are positive, and e is taken as positive when it is below the neutral axis):

A	area of concrete section
e	eccentricity of tendon
f_{ck}	28 day cylinder strength
f_{ctm}	mean tensile strength of concrete
M_{DL}	moment acting at transfer
M_F	moment under frequent loads
M_{QP}	moment from quasi-permanent loads
M_k	moment under characteristic loading
P_{m0}	prestress force at transfer $(t = 0)$
$P_{m\infty}$	final prestress force $(t = \infty)$
r_{inf}, r_{sup}	partial safety factors $(= 1.0$ in the UK)
Z_b	section modulus of bottom fibre
$Z_{b(25)}$	section modulus 25 mm below the tendon
Z_t	section modulus of top fibre

Lower chord, compression at transfer

$$\frac{1}{P_{m0}} \geq \frac{r_{sup}(1/A + e/Z_b)}{0.6f_{ck} + M_{DL}/Z_b} \quad \text{for} \quad r_{sup}P_{m0}\left(\frac{1}{A} - \frac{e}{Z_t}\right) + \frac{M_{DL}}{Z_t} \geq -f_{ctm}$$

Otherwise the calculation should be based on a cracked section.

Upper chord, tension at transfer
If the tensile stress at the upper chord exceeds f_{ctm}, check that the crack width does not exceed the appropriate value from Table 11.7. In this calculation, r_{sup} should be used in order to maximize the tensile stress, i.e. the prestressing force should be taken as $P = r_{sup}P_{m0}$.

Lower chord, tension at the serviceability limit state
Check decompression:

$$\frac{1}{P_{m\infty}} \le \frac{r_{inf}(1/A + e/Z_{b(25)})}{(M_F \text{ or } M_{QP})/Z_{b(25)}} \quad \text{for} \quad r_{inf}P_{m\infty}\left(\frac{1}{A}+\frac{e}{Z_b}\right) - \frac{M_k}{Z_b} \ge -f_{ctm}$$

Otherwise the designer should use a cracked section and adjust P and e until the neutral axis is 25 mm below the tendon.

Check the crack width: the crack width should be determined using r_{inf} in order to maximize tension, i.e. $P = r_{inf}P_{m\infty}$. The bending moment should be M_F for members with bonded tendons, or M_{QP} for members with unbonded tendons.

Check the tendon stress: calculate the tendon stress using a cracked section. The bending moment should be M_k. Both r_{sup} and r_{inf} should be considered, although the higher value is more likely to be critical.

Upper chord, compression at the serviceability limit state
Check that

$$\frac{1}{P_{m\infty}} \ge \frac{r_{inf}(1/A - e/Z_t)}{0.6f_{ck} - M_k/Z_t} \quad \text{for} \quad r_{inf}P_{m\infty}\left(\frac{1}{A}+\frac{e}{Z_b}\right) - \frac{M_k}{Z_b} \ge -f_{ctm}$$

$$\frac{1}{P_{m\infty}} \ge \frac{r_{inf}(1/A - e/Z_t)}{0.45f_{ck} - M_{QP}/Z_t} \quad \text{for} \quad r_{inf}P_{m\infty}\left(\frac{1}{A}+\frac{e}{Z_b}\right) - \frac{M_k}{Z_b} \ge -f_{ctm}$$

Otherwise, the calculation must be carried out on a cracked section.

Deformation
The span/depth ratios given in *clause 7.4.2* apply only to reinforced concrete. The designer must use the guidance given in *clause 7.4.3* when it is necessary to determine the deflections of prestressed concrete members.

Clause 7.4.2
Clause 7.4.3

11.6. Design of sections for shear and torsion
11.6.1. Shear
Ultimate limit state
In the design of sections for shear, EN 1992-1-1 follows the same procedure as for reinforced concrete, treating the prestressing force as an applied axial compression. The design shear force at a section should be calculated taking into account the external loads and the effect of any inclined tendons.

Clause 6.2.1(3)

Where the prestressing force increases the shear capacity, it should be multiplied by the partial safety factor $\gamma_{p,fav} = 1.0$ (0.9 in the UK); in other cases it should be multiplied by $\gamma_{p,unfav} = 1.3$ (1.1 in the UK). Because the prestressing force can occur on both the load and the resistance side of the equation, $\gamma_{p,fav}$ and $\gamma_{p,unfav}$ should be used consistently and the worst case adopted.

Clause 2.4.2.2(1)
Clause 2.4.2.2(2)

In members where the design shear force is less than the shear capacity of the concrete section alone, no design shear reinforcement is required, although the minimum shear reinforcement in accordance with *clause 9.2.2* should be provided unless the member is a slab or a lintel with a span ≤ 2 m.

Clause 6.2.1(3)
Clause 9.2.2

One of the factors governing the shear capacity of the concrete section alone is the area of longitudinal reinforcement in the tensile zone. EN 1992-1-1 does not state whether this

Clause 6.2.2(1)

should be the total area of both the normal reinforcement and the prestressing tendons or whether the areas should be adjusted to reflect the different design strengths. It is suggested that the sum of the areas be used for bonded tendons, as this term is included to account for the change in level of the neutral axis, which is governed more by reinforcement area than by its strength.

Clause 6.2.2(1)   The presence of prestress increases the shear capacity of the concrete section by 15% of the stress due to prestressing.

Clause 6.2.2(2)   For the special case of single span prestressed members without shear reinforcement, in zones where the maximum flexural tensile stress at the ultimate limit state is less than $f_{ctd} = f_{ctk0.05}/\gamma_c$, the shear capacity should be checked by limiting the maximum principal tensile stress in the cross-section to f_{ctd}. For pretensioned members, the effect of the prestress force is reduced in the transmission zone. The code draws the designer's attention to the fact that, in a member of varying width, the maximum principal tensile stress may not occur at the neutral axis. The section should be checked in several positions, and the lowest value for the shear capacity should be taken.

EN 1992-1-1 requires the width of the web used in determining the shear capacity by limiting the principal tensile stress to be reduced as follows:

Clause 6.2.3(6)
- Non-grouted ducts, grouted plastic ducts and unbonded tendons: $1.2 \times$ the sum of the duct diameters.
- Grouted metal ducts with a diameter $> 1/8$ of the web width: $0.5 \times$ the sum of the duct diameters.
- Grouted metal ducts with a diameter $\leq 1/8$ of the web width: 0.

Clause 6.2.2(1)   EN 1992-1-1 explains that the value of 1.2 is to avoid the splitting of the concrete compressive struts due to transverse tension, and suggests that it can be reduced to 1.0 if adequate transverse reinforcement is provided. It is not clear from the code that this reduction should also be applied when calculating the shear capacity of the concrete alone by the general method in *clause 6.2.2(1)*, but it is recommended that this should be the case where the ducts are located in the compression zone.

Clause 6.2.2(5)   The designer's attention is specifically drawn to the fact that, in regions cracked in flexure, the tensile force in the longitudinal reinforcement is increased by the presence of shear above that required for bending. Provision for this effect is made by shifting the bending moment diagram by a distance equal to the effective depth of the section so that the moment at any section is always increased.

Where shear reinforcement is required, it must be designed in accordance with the *Clause 6.2.3(3)* variable strut inclination method. The presence of prestress does not affect the area of shear reinforcement required, but it does increase the maximum shear capacity of the section, providing the mean prestress is less than $0.6f_{cd}$. This will be the case in most prestressed *Clause 6.2.3(6)* concrete members. In determining the maximum shear capacity of the section, the width of the concrete section taken into account should be reduced to allow for the presence of ducts, as described above.

Clause 6.2.3(7)   Again, the designer's attention is drawn to the fact that the tensile force in the longitudinal reinforcement is increased by the presence of shear above that required for bending. *Clause 9.2.1.3(2)* Provision for this effect is most easily made by shifting the bending moment diagram so that the moment at a given section is always increased. This is discussed in Chapter 10. As the angle of the struts with the horizontal is reduced, the area of shear reinforcement reduces, but there will be a corresponding increase in the longitudinal reinforcement requirement.

Clause 6.2.1(2)   The shear capacity of a member with shear reinforcement should be modified by taking into account the effects of the vertical components of the design force in inclined compression and tension chords. The effect of the inclined chords will be either positive or *Clause 6.2.1(6)* negative depending on the direction of inclination and the sign of the corresponding bending moment. The maximum shear capacity of a section reinforced in shear should be greater than the vector sum of the design shear force and the vertical components of the design forces in any inclined chords.

Serviceability

If shear reinforcement is detailed in accordance with the requirements discussed in Chapter 10, then it is not necessary to carry out any shear checks at the serviceability limit state.

11.6.2. Torsion

Torsional resistance must be calculated when the static equilibrium of the structure depends on its torsional resistance. EN 1992-1-1 states that the design should be carried out at both the serviceability and ultimate limit states; but design rules are only given for the ultimate limit state. The reinforcement is designed using the variable strut inclination method; the inclination of the struts must be the same as that used in designing for any coexistent shear forces. It can be assumed that serviceability requirements are met if the detailing rules discussed in Chapter 10 are followed.

Clause 6.3.1(1)P

Clause 6.3.2(2)

When torsion arises from consideration of compatibility only, EN 1992-1-1 requires the member to be designed to avoid excessive cracking. In practice, this means that stirrups and longitudinal reinforcement should satisfy specified detailing rules.

Clause 6.3.1(2)

11.7. Prestress losses

11.7.1. General

The designer's attention is drawn to the need to allow for losses of prestress when calculating the design forces in tendons at the various stages considered in the design. The causes of these losses are listed in EN 1992-1-1. Because of the uncertainty in estimating such losses, it is suggested that values are taken from European technical approval documents where these are available. In the absence of such data, EN 1992-1-1 suggests values for the various parameters that can be used for design.

Clause 5.10.3

11.7.2. Friction in jack and anchorages

Loss of prestressing force due to friction in the jack and anchorages is not mentioned in EN 1992-1-1. It is not generally a problem for the designer, because the design is normally based on the tendon force on the concrete member side of the jack. However, it does need to be allowed for in calibration of the actual jack used for stressing.

11.7.3. Duct friction

EN 1992-1-1 specifies the standard formula for calculating the force lost in overcoming duct friction when the tendon is stressed. This separates the loss into two parts:

Clause 5.10.5.2

(1) that caused by the tendon and duct following the specified profile
(2) that caused by slight variations in the actual line of the duct, which may cause additional points of contact between the tendon and the duct wall.

The losses are calculated in terms of the coefficient of friction μ and an unintentional angular displacement per metre length k. Values of μ for a range of internal tendons which fill about 50% of the duct and for a range of external unbonded tendons in different ducts are given in EN 1992-1-1, which also suggests that k should be taken between 0.005 and 0.01 rad/m. The loss of prestress due to unintentional angular displacements may be ignored for external tendons. Rather than using these values, it is generally better to refer to European technical approval documents or test results produced by the manufacturers of prestressing components; EN 1992-1-1 allows the designer to adopt this approach.

Clause 5.10.5.2(2)

Clause 5.10.5.2(3)
Clause 5.10.5.2(4)
Clause 5.10.5.2(1)

11.7.4. Elastic deformation

The method specified for the calculation of losses due to the immediate elastic deformation of the concrete when the prestressing force is applied follows normal procedures.

Clause 5.10.4
Clause 5.10.5.1

11.7.5. Anchorage draw-in or slip

EN 1992-1-1 mentions this cause of loss of prestressing force, but does not specify the values to be used. Appropriate values can be obtained from European technical approval documents and from the anchorage manufacturers, and should be checked on site, particularly if the member is short, when the loss due to this cause can be critical.

The loss due to anchorage draw-in, ΔP is given by

$$\Delta P = \frac{E_p A_{ps} \delta}{L} + mL$$

where

$$L = \sqrt{\frac{E_p A_{ps} \delta}{m}} \leq \text{length of tendon}$$

E_p is the modulus of elasticity of tendons
A_{ps} is the area of prestressing tendons
δ is the anchorage draw-in
m is the rate of change of prestressing force due to friction, which may be taken as $P_0 \mu(\alpha + k)$
μ is the coefficient of friction
α is the angular deviation per unit length
k is the unintentional angular deviation per unit length

11.7.6. Time-dependent losses

Clause 5.10.6
Time-dependent losses are those due to relaxation of the prestressing tendons and shrinkage and creep of the concrete. EN 1992-1-1 considers the loss due to each effect, as if that effect were acting alone, and then combines them taking 80% of the relaxation loss to account for the effect of creep and shrinkage reducing the stress in the tendon. The formula given in EN 1992-1-1 further reduces the combined loss to allow for the reduction in compressive stress in the concrete as the prestress force reduces with time.

Relaxation of steel

Clause 3.3.2(5)

Clause 3.3.2(6)
The starting point is the relaxation loss at 1000 h from an initial stress of 70% of the **actual** tensile strength of the tendon. This is best obtained from test certificates, or from the appropriate material standard. EN 1992-1-1 suggests values for this parameter in the absence of any other source of data. These values are numerically the same as those commonly taken as being the 1000 h relaxation loss from an initial stress of 70% of the **characteristic** strength of the tendon.

Table 11.8. Relaxation losses as a percentage of initial stress, σ_{pi}, at 20°C

Tendon type	Initial stress σ_{pi}	1	5	100	500	1000	5000	10 000	50 000	100 000	500 000
Class 1 (wire	$0.6f_{pk}$	0.3	0.5	1.2	2.0	2.4	3.9	4.8	7.8	9.6	15.5
or strand)	$0.7f_{pk}$	1.0	1.4	2.8	4.0	4.7	6.7	7.9	11.3	13.2	19.0
	$0.8f_{pk}$	3.3	4.1	6.5	8.3	9.2	11.7	13.0	16.5	18.3	23.3
Class 2 (wire	$0.6f_{pk}$	0.0	0.1	0.2	0.3	0.4	0.6	0.8	1.3	1.5	2.5
or strand)	$0.7f_{pk}$	0.2	0.3	0.6	0.8	1.0	1.4	1.6	2.3	2.7	3.9
	$0.8f_{pk}$	0.8	1.1	1.7	2.2	2.4	3.0	3.4	4.3	4.8	6.1
Class 3 (bars)	$0.6f_{pk}$	0.1	0.2	0.5	0.8	1.0	1.6	1.9	3.1	3.8	6.2
	$0.7f_{pk}$	0.5	0.7	1.3	1.8	2.1	3.1	3.6	5.2	6.0	8.7
	$0.8f_{pk}$	1.7	2.2	3.4	4.3	4.8	6.1	6.7	8.6	9.5	12.1

Table 11.9. Factor to allow for non-linear creep

$\sigma_c/f_{cm}(t_0)$	Factor
≤ 0.45	1.00
0.50	1.08
0.55	1.16
0.60	1.25
0.65	1.35
0.70	1.46
0.75	1.57

The 1000 h relaxation loss is multiplied by a factor which depends on the relaxation class of the tendon, the ratio of the initial stress in the tendon to its **characteristic** tensile strength and duration. A duration of 500 000 h can be used to determine the final (long-term) relaxation loss. Relaxation losses are given in Table 11.8.   *Clause 3.3.2(7)*
   Clause 3.3.2(8)

The values for the relaxation losses in Table 11.8 vary from those that the designer might have expected from previous experience. In particular, the values for the loss at 1000 h are less than those which would normally have been previously assumed. This can be explained by the fact that the losses in the table are for initial stresses expressed as a function of the **characteristic** tendon strength. Actual tendon strengths can be up to 14% higher than the characteristic strength, so that an initial stress of 70% of the **actual** tendon strength could be numerically equal to an initial stress of $70 \times 1.14 \approx 80\%$ of the **characteristic** strength. For this value of initial stress it can be seen that the 1000 h loss is approximately equal to the suggested values, and the formulae in EN 1992-1-1 can be used with confidence.

Shrinkage

EN 1992-1-1 suggests typical values within the main text for the final shrinkage strain of concrete (*Tables 3.2 and 3.3*), with more detailed information in *Annex B*.   *Clause 3.1.4(6)*

Creep

EN 1992-1-1 suggests typical values within the main text for the final specific creep strain (or creep coefficient), $\phi(\infty, t_0)$, for concrete (*Fig. 3.1*), with more detailed information in *Annex B*. The designer should note that the creep coefficient is related to the tangent modulus of the concrete, E_c, which may be taken as $1.05E_{cm}$. The creep deformation under a constant compressive stress, σ_c, is given by   *Clause 3.1.4(2)*

   Clause 3.1.4(3)

$$\varepsilon_{cc}(\infty, t_0) = \phi(\infty, t_0)\frac{\sigma_c}{1.05E_{cm}}$$

When the compressive stress at the time of loading exceeds $0.45f_{ck}(t_0)$, then the creep deformations can develop non-linearly. The non-linear creep coefficient, $\phi_k(\infty, t_0)$, is determined by multiplying the linear creep coefficient, $\phi(\infty, t_0)$, by a factor obtained from Table 11.9.   *Clause 3.1.4(4)*

11.8. Anchorage zones

11.8.1. Pretensioned members

Guidance is provided in EN 1992-1-1 for estimating the transmission lengths in pretensioned members. A distinction is drawn between the transmission length over which the prestressing force is fully transmitted to the concrete l_{pt}; the dispersion length over which the concrete stresses gradually disperse to a distribution compatible with plane sections remaining plane l_{disp}; and the anchorage length over which the tendon force at the ultimate limit state is fully transmitted to the concrete l_{bpd}. The designer will find that this distinction   *Clause 8.10.2*
   Clause 8.10.2.1

Clause 8.10.2.2
Clause 8.10.2.3
Clause 8.10.2.3(1) helps to clarify the distribution of forces within the anchorage zone. Typical transmission and anchorage lengths for good bond conditions are given in Table 11.10.

The anchorage of pretensioned tendons at the ultimate limit state need only be checked if the concrete tensile stress in the anchorage zone exceeds $f_{ct0.05k}$. The tendon force to be used should be calculated assuming a cracked cross-section and include the effect of shear.

Often pretensioned tendons are well distributed across the cross-section of the member so that forces induced by transfer of the prestressing force from the tendon positions to a linear distribution are low. In slabs these forces can often be resisted by the concrete alone without the need for ordinary reinforcement, whilst in beams it is generally satisfactory to reduce the stirrup spacing to approximately 75 mm throughout the transmission length. Where this is not the case, reinforcement to maintain overall equilibrium and carry any spalling forces should be provided as described in Section 11.8.2.

11.8.2. Post-tensioned members

The design of anchorage zones, or end blocks, in post-tensioned members is a very important
Clause 8.10.3 area of design, but one that is rarely well-treated in codes of practice. EN 1992-1-1 is no exception, although it does draw attention to this important area and outlines the principles of the design.

The designer's attention is drawn to the need to consider bearing stresses behind the anchorage, overall equilibrium and transverse tensile forces, i.e. bursting and spalling forces.
Clause 2.4.2.2(3)
Clause 8.10.3(4)
Clause 8.10.3(3) The design is carried out at the ultimate limit state, and is based on γ_p (= 1.2) times the prestressing force. No check of crack widths is required when reinforcement stresses are limited to 300 MPa. Bearing stresses under anchorage plates should be checked in accordance with the relevant European technical approval. In practice, this means that the designer should check with the supplier the minimum concrete strength and reinforcement
Clause 8.10.3(4)
Clause 8.10.3(5)
Clause 6.5.3(3) required adjacent to the anchorage. It is suggested that overall equilibrium be checked using a strut-and-tie method, and guidance is given on this and on the angle of dispersion of the prestressing force into the concrete section. Formulae are also given for determining the bursting forces behind an anchorage. This is of benefit to the designer. However, no guidance is given on how to design against spalling forces.

The designer should give particular attention to anchorage zones that have a cross-section of different shape from that of the general cross-section of the member.

For more information on the design and detailing of anchorage zones in prestressed concrete, the designer is referred to the excellent documents produced by CIRIA[14] and the Institution of Structural Engineers.[15]

Table 11.10. Transmission lengths, l_{pt}, and anchorage lengths, l_{bpd}, for good bond conditions: number of diameters

Tendon type	Concrete grade					
	25/30	30/37	35/45	40/50	45/55	50/60
7-wire strands						
l_{pt}	60–90	50–80	45–70	40–65	40–60	35–55
l_{bpd}	130	115	100	95	85	80
Indented wire						
l_{pt}	90–135	80–120	75–110	65–100	60–90	55–85
l_{bpd}	170	150	135	125	115	110

Ultimate stress in tendon = 1304 MPa.
Stress in tendon immediately after release = 1250 MPa.
Stress in tendon after all losses = 875 MPa.

Table 11.11. Minimum clear spacing requirements for prestressing tendons and ducts

Pretensioned	Vertically	$\geq d_g$	$\geq 2\phi$
	Horizontally	$\geq d_g + 5$ mm	$\geq 2\phi \geq 20$ mm
Post-tensioned	Vertically	$\geq d_g$	$\geq \phi \geq 40$ mm
	Horizontally	$\geq d_g + 5$ mm	$\geq \phi \geq 50$ mm

ϕ is the diameter of a tendon or duct, as appropriate; d_g is the aggregate size.

11.9. Detailing

The detailing provisions given in *Sections 8* and *9* of EN 1992-1-1 apply to both prestressed and normally reinforced concrete. In this section, only provisions that specifically relate to prestressed concrete are discussed. A more general commentary on the detailing requirements of EN 1992-1-1 is given in Chapter 10.

11.9.1. Spacing of tendons and ducts

EN 1992-1-1 specifies the minimum allowable clear spacing between adjacent tendons or ducts, which is generally the greatest of twice the tendon diameter, the duct diameter or the aggregate size plus 5 mm. Absolute minimum clear spacings are also specified, but these are unlikely to be critical when account is taken of the need to be able to compact the concrete around the tendons. For post-tensioned members, EN 1992-1-1 bases its requirements on the outside dimensions of the duct rather than on the internal dimensions, the latter being more normal UK practice. The requirements are summarized in Table 11.11. *Clause 8.10.1.2*

Clause 8.10.1.3

Clause 8.10.1.1

EN 1992-1-1 does not specify any additional requirements for the minimum clear spacing in the plane of curvature between curved tendons. This is a serious omission, as very high local forces can be developed, particularly when tendons in outer ducts are stressed before the inner ducts are grouted. The designer is recommended to use the requirements given in other codes, such as BS 8110, which are reproduced in Table 11.12.

11.9.2. Anchorages and couplers

EN 1992-1-1 requires that not more than 50% of the tendons be coupled at any one cross-section unless it can be shown that a higher percentage will not cause more risk to the safety of the structure. This is contrary to the generally accepted UK practice of coupling all the tendons at one cross-section. The reason for the EN 1992-1-1 requirement is that when a coupled tendon is stressed, the force between the previously stressed concrete and the anchorage decreases, and the local deformation of this concrete is reduced. Increased compressive forces are induced adjacent to the anchorage, and balancing tensile forces between adjacent anchorages. Uncoupled tendons across the joint reduce this effect, as well as assisting in carrying the induced tensile forces. Alternatively, it is suggested that such forces could be carried by properly detailed normal reinforcement. *Clause 8.10.4(5)*

11.9.3. Minimum area of tendons

EN 1992-1-1 specifies minimum areas of reinforcement to avoid brittle failure and wide cracks occurring and also to resist forces arising from restrained actions. In prestressed members, minimum reinforcement need not be provided to meet this requirement where one of the following conditions is satisfied: *Clause 9.1*

Clause 7.3.2(4)

(1) the maximum tensile stress in the concrete at serviceability under the characteristic load combination is less than f_{ctm} or
(2) the prestress and ordinary reinforcement is sufficient to limit crack widths to the required values

and brittle failure is avoided by: *Clause 5.10.1(6)*

Table 11.2. Minimum distance between centre-lines of ducts in the plane of curvature

Minimum distance (mm)

Radius of curvature of duct (m)	Duct internal diameter (mm): 19	30	40	50	60	70	80	90	100	110	120	130	140	150	160	170
	Tendon force (kN): 296	387	960	1337	1920	2640	3360	4320	5183	6019	7200	8640	9424	10338	11248	13200
2	110	140	350	485	700	960										
4	55	70	175	245	350	480	610	785	940					Radii not		
6	38	60	120	165	235	320	410	525	630	730	870	1045		normally		
8			90	125	175	240	305	395	470	545	655	785	855	used 940		
10			80	100	140	195	245	315	375	440	525	630	685	750	815	
12						160	205	265	315	365	435	525	570	625	680	800
14						140	175	225	270	315	375	450	490	535	585	785
16							160	195	235	275	330	395	430	470	510	600
18								180	210	245	290	350	380	420	455	535
20									200	220	265	315	345	375	410	480
22											240	285	310	340	370	435
24												265	285	315	340	400
26												260	280	300	320	370
28																345
30																340
32																
34																
36																
38																
40	38	60	80	100	120	140	160	180	200	220	240	260	280	300	320	340

(1) The tendon force shown is the maximum normally available for the given size of duct (taken as 80% of the characteristic strength of the tendon).

(2) Values less than 2 × duct internal diameters are not included.

(3) Where tendon profilers or spacers are provided in the ducts and these are of a type which will concentrate the radial force, the values given in the table will need to be increased. If necessary, reinforcement should be provided between ducts.

(4) The distance for a given combination of duct internal diameter and radius of curvature shown in the table may be reduced in proportion to the tendon force when this is less than the value tabulated, subject to meeting the minimum requirements of Table 11.1.

(1) the member containing pretensioned tendons
(2) access being provided to enable non-destructive testing and monitoring of the condition of the tendons to be carried out easily
(3) satisfactory evidence being provided on the reliability of the tendons or
(4) proportioning the section to ensure that, under the frequent combination of loads, cracking occurs before the ultimate bending capacity is reached.

For members with unbonded or external tendons, EN 1992-1-1 requires that the ultimate bending capacity exceeds the cracking moment by at least 15%. This seems an appropriate value to adopt also for members with bonded tendons.

Clause 9.2.1.1(4)

11.9.4. Tendon profiles

The design for shear in EN 1992-1-1 leads to a requirement for longitudinal reinforcement in addition to that required for bending. This reinforcement is generally allowed for by a horizontal displacement of the bending moment envelope (the 'shift' rule). When the flexural design of a prestressed concrete member is governed by the ultimate limit state, which could be the case when the design criterion is a limiting crack width, the longitudinal reinforcement will need to be increased and/or the tendon profile adjusted in order to resist the tensile forces arising from the effect of inclined cracks in the webs.

Clause 9.2.1.3

CHAPTER 12

Structural fire design

12.1. Aims of design

Structural fire design is covered in EN 1991-1-2. Under fire conditions, a structure will be required to fulfil load bearing and/or separating function. The code provides guidance for satisfying the above requirements and this chapter of the manual discusses the salient features. Building regulations normally specify the fire rating for the building. It can also be calculated from first principles using EN 1991-1-2 (*Actions on Structures Exposed to Fire*).

The structure is required to comply with criteria R, I and E as appropriate:

- Criterion R is assumed to be satisfied if the structure maintains adequate mechanical resistance for the required duration of fire exposure.
- Criterion I is the insulation criterion, which also applies to members fulfilling separating function. It is deemed to be satisfied if the average rise in temperature on the non-exposed surface does not exceed 140 K and the maximum rise does not exceed 180 K.
- Criterion E applies to elements required to fulfil separating function, and will be satisfied if the integrity of the structure is maintained for the required duration such that the spread of flames, hot gases and excessive heat is avoided.

12.2. Design procedure

The designer has a wide choice of methods. The method for verifying the fire resistance in a project should be chosen on the basis of acceptability to regulatory authorities. The design can be based on thermal actions given by either nominal fire exposure or those calculated using the physical parameters of the building. The former is the most common method, and the latter should only be undertaken by designers with adequate appreciation of all aspects of fire engineering.

In both procedures it is further permitted to base the design on (1) member analysis (2) analysis of parts of the structure or (3) analysis of the whole structure.

In member analysis – which includes tabulated data – only the effects of thermal gradients across the cross-section is considered; the effects of axial or in plane thermal expansions are neglected. Also, the boundary conditions are assumed to remain unchanged throughout the fire exposure.

In the analysis of a part of the structure, a section of the structure is isolated such that stationary boundary conditions could be prescribed in the analysis. Within the part that is isolated, the relevant failure mode in fire exposure, temperature-dependent material properties and the effects of thermal expansions and deformations should be taken into account in the analysis.

In the analysis of the whole structure, none of the approximations noted above are used, and the global analysis is undertaken accounting for the temperature-dependent material properties, member stiffness and the effects of thermal expansions and deformations.

The code gives only limited guidance for methods 2 and 3 above.

12.3. Actions and partial factors

12.3.1. Actions

Where calculations are undertaken, the load combination to be used in fire conditions is given in EN 1990 (*Basis of Structural Design*). It is

$$G_k + A_d + \psi_{1,1} Q_{k,1} + \psi_{2,i} Q_{k,i}$$

where A_d is the design value of the accidental action (in this case actions caused by fire), $Q_{k,1}$ is the leading variable action and $Q_{k,i}$ represents other accompanying actions, if applicable.

For member analysis, where the effects of thermal deformation are neglected and where there is only one variable action, the action effect under fire conditions may be taken as η_{fi} times the value of action effect used in normal temperature design. The value of η_{fi} depends on the ratio of variable and permanent loads and the use of the building. EN 1992-1-2 provides a figure to arrive at the value. A value of 0.7 may be used as a safe approximation in all cases.

12.3.2. Material factors

EN 1992-1-2 recommends a value of 1.0 for both γ_C and γ_S. The UK National Annex has also decided to adopt these values. The reasoning is as follows.

In probabilistic terms

$$\gamma_{M,fi} = \exp(0.7 V_R - 1.645 V_f)$$

where

$$V_R = (V_m^2 + V_G^2 + V_f^2)^{0.5}$$

V_m is the coefficient of variation for model uncertainty, V_G is the coefficient of variation for geometry of element and V_f is the coefficient of variation for material property. Values for V_f and V_G were taken as the same as for normal design at room temperature, and $V_{m,fi}$ was taken as $2V_{m,\text{room temperature}}$. This resulted in $\gamma_c = 0.84$ and $\gamma_s = 1.0$. It was decided to use the value of 1.0 for all material properties.

12.4. Member analysis using tabular data

12.4.1. Scope

Section 5 of EN 1992-1-2 sets out minimum dimensions and axis distances (cover plus half the diameter of the reinforcement bar) for different types of structural element for fire exposure periods of 30, 60, 90, 120, 180 and 240 min. Some detailing requirements are also given. The member types covered are columns, load-bearing and non-load-bearing walls, beams (simply supported and continuous) and slabs (simply supported and continuous boundary conditions, one- and two-way spanning slabs, solid and ribbed slabs, and flat slabs).

The table values apply to concrete made with siliceous aggregates and with a density in the range 2000–2600 kg/m³. If calcareous or lightweight aggregate is used, the table values for minimum dimensions for slabs and beams may be reduced by 10%.

When using tabulated data, no further checks are required for shear, torsion, anchorage details and explosive spalling of concrete (spalling related to the moisture content of the concrete). However, the code requires surface reinforcement when the axis distance in a member exceeds 70 mm to prevent falling off of the cover concrete.

Rather than repeating the data tabulated in the code here, this guide discusses the basis of the tables and some of the limitations.

12.4.2. Basis for the tabulated data

The tables have been developed on an empirical basis, general experience in Europe and theoretical evaluation of tests. In particular, the data for method A for columns are based on recent tests in Belgium and Canada. The data for the more common types of structural elements are derived making approximate and conservative assumptions.

When the thickness of walls and slabs complies with the minimum dimensions, the integrity and insulation criteria (see Section 12.4.1 above for definition) for the separating function may be assumed to be satisfied.

For the load-bearing function, the following assumptions apply:

- $E_{d, fi}/R_{d, fi} \leq 1.0$, where $E_{d, fi}$ is the design effect of actions under fire conditions and $R_{d, fi}$ is the design resistance of the element under fire conditions.
- Unless otherwise noted, it is assumed that $\eta_{fi} = E_{d, fi}/E_d = 0.7$, where E_d is the design effect of actions in normal temperature design.
- The critical temperature of reinforcement (θ_{cr}) (i.e. the temperature of reinforcement at which the failure of the member is expected to occur under fire conditions for a given level of steel stress) is 500°C. This assumes that the stress level $= \sigma_{s, fi}/f_{yk} = 0.6$, $\eta_{fi} = 0.7$ and $\gamma_s = 1.15$.
- For prestressing tendons, $\theta_{cr} = 400°C$, and for prestressing wires and strands $\theta_{cr} = 350°C$. The stress level for both is taken as 0.55.

The thickness of slabs in the tables is to ensure adequacy for the separating function, and may include the thickness of any non-combustible floor finishes in addition to the thickness of the structural floor.

The axis distance in the tables should be compared with (1) the axis distance to a single layer of reinforcement of the same diameter; or (2) the weighted axis distance of bars in the case of reinforcement in multiple layers or multiple sizes.

12.4.3. Discussion of some features

Columns

Two alternatives are provided. Both methods apply to braced structures. No guidance is given for unbraced frames.

Method A is based on tests, and has been derived by curve fitting. For different fire resistance periods and load levels, it provides the minimum dimension of columns and the axis distance to the main reinforcement. The following limitations to the validity are imposed to reflect the limits of the test data:

- The effective height under fire conditions is $l_{0, fi} \leq 3$ m. Under fire conditions, $l_{0, fi}$ may be taken as $0.5l$ for intermediate floors and between $0.5l$ and $0.7l$ for columns in top storeys of buildings. In the above, l is the distance between the centres of lateral restraints. The effective lengths might seem small at first sight, but it should be appreciated that the effective length depends on the stiffness (EI) of members meeting at a joint. Due to the action of fire, the value of E will be considerably lower in the members in the storey in which the fire occurs compared with those in the floors above and below.
- The code limits the first-order eccentricity to that used in the tests. Under fire conditions $e = M_{0Ed, fi}/N_{0Ed, fi}$ should not exceed e_{max}. The code recommends a value of $0.15h$ (or b) but permits a range of between $0.15h$ (or b) and $0.40h$ (or b). The value of the eccentricity e under fire conditions may be taken to be the same as in normal temperature design.

Data are given for different load levels $\mu_{fi} = N_{Ed, fi}/N_{Rd}$, where $N_{Ed, fi}$ is the load in a fire situation and N_{Rd} is the design resistance under normal temperature conditions, calculated using EN 1992-1-1.

The code also gives a formula for arriving at the fire resistance period when the parameters are different to those used for the table. This includes load level, α_{cc}, axis distance, number of bars and effective length of the column. The value calculated should be greater than the actual fire period required.

The alternative method, B, is a subset of *Annex C*, which has been derived by buckling analysis as set out in *Annex B3*. It compares reasonably well with the results of tests although it is more conservative compared with method A. This method gives data similar to that in method A, but another variable, the mechanical reinforcement ratio (ω), has been brought into the table. The limits of validity are as follows:

- The first-order eccentricity e under fire conditions should not exceed $0.025b$ or 100 mm.
- The slenderness ratio $\lambda_{fi} = (l_{0,fi}/i)$ should not exceed 30. The effective length of columns may be taken as that defined for method A.
- The load level $n = N_{0Ed,fi}/0.7(A_c f_{cd} + A_s f_{yd})$. $N_{0Ed,fi}$ may be taken as $\eta_{fi}N_{0Ed}$, and η_{fi} may be taken conservatively as 0.7.

Annex C comprises nine tables grouped as follows: *Tables C1, C2* and *C3* apply to $\omega = 0.1$ and $e = 0.025b$, $0.25b$ and $0.5b$, respectively; *Tables C4, C5* and *C6* deal with $\omega = 0.5b$ and $e = 0.025b$, $0.25b$ and $0.5b$, respectively; and *Tables C7, C8* and *C9* are for $\omega = 1.0$ and $e = 0.025b$, $0.25b$ and $0.5b$, respectively. In each table, the minimum size of columns and the axis distance to the main bars are given for different fire periods, load levels and slenderness ratios.

Beams

In continuous beams it should be noted that there are requirements for top reinforcement over supports and beam width over the first intermediate support. In order to use the data for continuous beams, the above requirements should be complied with, and redistribution of bending moments for normal temperature design should not exceed 15%. Otherwise the beams should be treated as simply supported for fire design.

Slabs

The data for two-way spanning slabs are applicable only to slabs supported on all four edges. If this is not met, the slab should be treated as one-way spanning.

The thickness of the slab for continuous slabs is the same as for simply supported slabs. But some reduction in the axis distance will apply to continuous slabs if the redistribution of bending moments for normal-temperature design does not exceed 15% and top reinforcement over intermediate supports complies with the requirements mentioned for beams.

There is also a requirement to provide top reinforcement of at least 0.5% A_c over intermediate supports if any of the following apply: (1) use of cold worked reinforcement; (2) the end supports in a continuous slab do not have fixity or are not detailed in accordance with EN 1992-1-1; or (3) the load effects cannot be distributed laterally.

In flat slabs, redistribution of bending moment for normal-temperature design should not exceed 15%. For fire ratings of 90 min and above, top reinforcement should be provided over the full span in the column strips. The area of reinforcement should be at least 20% of the total top reinforcement required over intermediate supports.

12.5. Simple calculation methods

Three calculation methods, which will have considerable application in practice, are noted in *Annexes B* and *E*.

12.5.1. 500°C isotherm method

This method requires knowledge of the temperature distribution across the cross-section of the member. EN 1992-1-2 provides such profiles for slabs, beams and columns in *Annex A* for standard time–temperature relationships. The method may be applied to sections, the minimum width of the cross-section of which complies with the values in *Table B1*.

The thickness of the zone damaged by heat is first identified. This is taken as the average depth of the 500°C isotherm in the compression zone of the cross-section. This damaged concrete is discounted and assumed not to contribute to the load-bearing capacity. The remaining cross-section is assumed to retain the initial values for strength and modulus of elasticity.

Although some reinforcement bars may lie outside the reduced cross-section, they are still taken into account in the calculations. By applying the reduction factors appropriate to their temperature, the strength of all reinforcement is reduced.

With the above assumptions, the problem reduces to the calculation of the resistance of the cross-section using the methods used for normal-temperature design.

12.5.2. Zone method
This method is more laborious, but more accurate, than the 500°C isotherm method.

The first step in the calculation is the estimation of the thickness a_z of fire damaged zone. This is carried out for an equivalent wall exposed to fire on both sides. The thickness of the equivalent wall is the actual thickness in the case of members subject to fire on opposite faces (e.g. walls, webs of beams and rectangular columns). The thickness is twice the actual thickness in the case of members subject to fire on one face only (e.g. slabs). The procedure then is as follows:

- Divide the half thickness of the equivalent wall into a number (minimum of three) of parallel zones.
- Using *Fig. B5a* or figures in *Section 4* of EN 1992-1-2 in conjunction with the temperature profile in *Annex A*, estimate the reduction factor $(k_c(\theta_i))$ at the centre of each zone and on the centreline of the equivalent wall $(k_c(\theta_M))$. The code provides formulae for the calculation of the mean reduction factor for half the width of the equivalent wall and the thickness of the damaged section a_z.
- Proceed with the calculation of the resistance with the reduced cross-section as in the 500°C isotherm method.

Although not mentioned in the code, it is the authors' opinion that as an alternative to the above method, the following procedure can be used:

- Discount 25 mm from the faces of the members exposed to fire.
- using the temperature profiles in *Annex A*, estimate the temperature at the level of reinforcement and in a number of layers of concrete.
- It should now be possible to calculate the resistance of the cross-sections from first principles.

12.5.3. *Annex E* method
This is a useful annex for the calculation of the resistance of beams and slabs subject to fire. Guidance is given for both simply supported and continuous members.

References

1. Schlaich, J. and Schafer, K. (1991) Design and detailing of structural concrete using strut-and-tie models. *Structural Engineer*, **69**, 113–125
2. Johansen, K. W. (1962) *Yield Line Theory*. Cement and Concrete Association, London
3. CEB Commission 'Dalles et Planchers' (1962) The application of yield line theory for the calculation of the flexural strength of slabs and flat slab floors. *CEB Bulletin 35*. FIB, Lausanne.
4. Wood, R. H. and Armer, G. S. T (1970) *The Strip Method for Designing Slabs*. BRE, Watford, Current Paper 39/70.
5. Whittle, R. T. (1994) *Design of Reinforced Concrete Flat Slabs to BS 8110. CIRIA Report 110*, revised edn. CIRIA, London.
6. British Standards Institute (2005) *Background Paper for the UK National Annex to BS EN 1992-1-1*. BSI, London, BS PD 6687.
7. Comité Euro-International du Beton (1978) *CEB/FIP Model Code for Concrete Structures*. FIB, Lausanne.
8. Comité Euro-International du Beton (1982) *CEB/FIP Manual on Bending and Compression. CEB Bulletin d'Information 141*. FIB, Lausanne.
9. Comité Euro-International du Beton (1988) *Manual for the Design of Durable Structures. CEB Bulletin d'Information 185*. FIB, Lausanne.
10. Rowe, R. E. (1987) Handbook to British Standard BS 8110:1985: Structural Use of Concrete. Palladian, London.
11. Beeby, A.W. (1978) *Cracking and Corrosion. Concrete in the Oceans Report No. 2*. CIRIA Underwater Engineering Group, London.
12. Schiessl, P. and Raupach, M. (1997) Laboratory studies and calculations on the influence of chloride-induced corrosion of steel in concrete. *ACI Materials Journal*, **94**, 56–62.
13. Comité Euro-International du Beton (1989) *CEB Design Guide on Durable Concrete Structures*, 2nd edn. *CEB Bulletin d'Information 182*. FIB, Lausanne.
16. Clarke, J. L. (1976) *A Guide to the Design of Anchor Blocks for Post-tensioned Prestressed Concrete. Guide No. 1*. CIRIA, London.
17. Institution of Structural Engineers (2006) *Manual for the Design of Concrete Building Structures to Eurocode 2*. Institution of Structural Engineers, London.

Index